The Geography of Context

The Geography of Context

Nicholas Fotion

Hamilton Books
Lanham • Boulder • New York • London

The Rowman & Littlefield Publishing Group, Inc.
An imprint of The Rowman & Littlefield Publishing Group, Inc.
4501 Forbes Boulevard, Suite 200, Lanham, Maryland 20706
Hamilton Books Acquisitions Department (301) 459-3366

6 Tinworth Street, London SE11 5AL

Copyright © 2019 by The Rowman & Littlefield Publishing Group, Inc.

All rights reserved. No part of this book may be reproduced in any form or by any electronic or mechanical means, including information storage and retrieval systems, without written permission from the publisher, except by a reviewer who may quote passages in a review.

British Library Cataloguing in Publication Information Available

Library of Congress Cataloging-in-Publication Data

Library of Congress Control Number: 2019930031
ISBN: 9780761871033 (pbk. : alk. cloth) | ISBN: 9780761871040 (electronic)

∞™ The paper used in this publication meets the minimum requirements of American National Standard for Information Sciences Permanence of Paper for Printed Library Materials, ANSI/NISO Z39.48-1992.

Printed in the United States of America

Contents

Introduction vii

1 Text and Context 1
2 More Surface Material 9
3 Social Contexts 15
4 Ethical (Moral) Memories 21
5 Empirical Contexts 27
6 More Empirical and Normative Contexts 35
7 "Logic" and Contexts 41
8 Still More "Logic" 49
9 Expressives and Declarations 57
10 Putting it Together 65
11 Four Processes 75
12 Interpretation 87
13 Closing Thoughts 97

Bibliography 105
Index 107
About the Author 109

Introduction

This is an essay about context. It is *not* a doctoral-type document burdened with endless collections of endnotes. Like a painting, an essay invites careful examination of what it presents. As such, it is *not* especially concerned to discuss the views of others on the topic of linguistic context.[1] These views tell us much of what we want to know about how language works. However, certain parts of linguistic context are not well explored by these views. This essay, then, aims to present a portrait of what is, relatively speaking, unexplored. One promising way to explore these parts of context is to return to the study of speech acts. Speech act theory had its hey-day between the 1950s into the 1980s.[2] After that, philosophers and linguists turned to other matters. But in turning away, speech act theorists and others failed to appreciate the full power of the theory. So this essay engages in a program of speech-act appreciation and thereby helps us appreciate aspects of linguistic context more fully.

The main reason speech act theory is useful in studying linguistic context is that the theory is multi-faceted. It does not privilege true-and-false, so called assertive, claims as many philosophers, logicians, and linguists did, and still do. The theory argues that meaningful speech acts include assertives, but also a wide variety of other acts, each of which has its own set of rules or standards. In being multi-faceted, the theory reflects the multi-faceted nature of the concept of context.

The father of speech act theory is John Austin. His lectures at Harvard in 1955 introduced as series of concepts that later all fell under the heading of speech act theory.[3] If John Austin is the father, John Searle is the son. Searle took over the burden of developing and refining the theory when Austin died prematurely in 1966. In a small book titled *How to Do Things With Words* that emerged from his lectures, Austin attempted to classify the hundreds of different types of speech acts (performatives) into five main categories. Later, in 1979, Searle reworked and systematized Austin's categories of speech into five different categories. In this work, I follow Searle with one major modification. Here, then, are Searle's five major types of speech acts.[4] What follows are rough-and-ready characterizations that will receive more clarification in later chapters.

- ASSERTIVES (Austin called them constatives). With these speech acts, the speaker (writer, signaler) informs the hearer (reader, etc.)

or hearers that what has been said is either true or false. Like the other major speech acts, there are many sub-types that fall under the assertive heading. These include reports of facts, hypotheses, guesses, and predictions.
- COMMISSIVES. These speech acts bind the speaker to perform some action on behalf of the hearer or for the one(s) designated by the hearer. Promises, vows, contract signings, pledges of allegiance, etc. represent some of the sub-categories.
- DIRECTIVES. Hearers here are bid by the speaker to perform some action or actions. Falling under this heading are orders, requests, pleadings, etc.
- EXPRESSIVES. These speech acts fall under two headings. Some are social lubricants. Examples are thankings, greetings, condolences, and congratulations. So-called social irritants constitute the second heading. Included here are a group of colorful expressions that one is reluctant to put in print. A printable example is "You're a dirty dog."
- DECLARATIONS. When these speech acts are issued, the hearers "magically" change their social status. Hiring applicants and later firing them are examples. So are promotions, as are certain things said at degradation ceremonies (e.g., in the military). This class of speech acts will prove to be important in this study.
- EVALUATIVES. These are the speech acts that I add to Searle's list. These speech acts rank things, events, and persons as better or worse. "She is an A student" is an example.

The aim of this study is to describe certain important aspects of linguistic context. It is as if these aspects of context represent a planet, and that it is a mistake to pay too much attention just to one of its continents. It would be better to give serious attention to all of them since what is true about the one we focus on most may have features not found in the other continents and/or may lack features found elsewhere. Indeed, I will claim, the "continents" of context are surprisingly varied, so much so that one is hard pressed to uncover one "continent" that we can call typical.

In this "geography" lesson, I will describe one "continent" or domain of context at a time. Once that task is over, I will cautiously draw some generalizations about context. These generalizations probably fall short of growing into a theory or model. It may be that the variety found in one "continent," and then another, may be too great to allow for theory formation.

NOTES

1. Richard Montague, *Formal Philosophy: Selected Papers of Richard Montague*, edited by R. Thomason (New Haven: Yale University Press, 1974); David Kaplan, "Demon-

stratives" in J. Almog, etc. editors *Themes from Kaplan* (New York and Oxford: Oxford University Press, 1989), pp. 481–563; David Lewis, *Papers in Philosophical Logic* (Cambridge: Cambridge University Press, 1998); Robert Stalnaker, *Context* (Oxford: Oxford University Press, 2014).

2. Kent Bach and Robert M. Harnish, *Linguistic Communications and Speech Acts*, Cambridge, MA and London: The MIT Press, 1979. Second Printing 1980.

3. J. L. Austin, *How to Do Things With Words*, 2nd edition, Cambridge, MA: Harvard University Press, 1962 and 1975. Edited by J. O. Urmson and Marina Sbisà.

4. John R. Searle, "A taxonomy of illocutionary acts," in *Expression and Meaning*, Cambridge: Cambridge University Press, 1979, 1–29.

ONE
Text and Context

TEXTS, MASTER SPEECH ACTS, AND SURFACE CONTEXTS

Studying (linguistic) context is a bit like studying the material that makes our planet what it is. With our planet, the material is layered many times over. Also, as one stands on the earth, one can't help noticing how the layer of the planet just below one's feet changes with the weather. If standing in the desert, change is visible moment to moment. Deeper down changes are less frequent.

    It is that way with linguistic context. It too is layered many times over and it too changes on the surface more quickly than "down deep." It might be thought, then, that the place to start studying would be "down deep." Being more stable, it is likely to be more easily understood. But, sadly, as with the study of the planetary stuff, deep-down studies are hard to pull off. Digging is hard work. So we dig down deep if we must. But the ease of reaching and studying surface phenomena tempts us to start our work on that level. It is the same with the study of linguistic contexts. It is easier to start at the surface and gradually work down to where all that deep stuff resides. That is where this study begins: surface contexts first, just below the surface second, the layer below third, and so on.

    Before scratching the surface, there is a need to make a distinction, crudely at first, between what constitutes context and what does not. I will use the term "text" to stand for what is not part of the context. In a conversation between Billy and Charley about Derek Jeter's last year in baseball (2014),[1] what Billy says now about Jeter is part of their conversational text. Charley's just prior comments can, but need not, also be taken as part of the text. As are Billy's comments on Charley's just prior comments. Charley's immediate future comments on what Billy said just now can also be taken as part of the text. How far back we can go to say that we are looking at their texts is not clear. However, it is tempting to say that if their conversation is both continuous and relatively brief, what both of them say can be taken as part of the text. If, on the next day, they continue their Jeter-conversation, one is tempted, instead, to classify what they said yesterday as part of the context instead of the text. What they said back then is presupposed in their next-day's conversation.

If, on another day when they meet, Billy says to Charley "Let's talk about the Middle East situation," Billy has issued what I will call a master speech act.[2] A master speech act in one sense is an ordinary speech act. Following Searle's classification system of basic speech acts, Billy's suggestion is a directive and also a commissive. It directs Charley to talk about the Middle East; and it (conditionally) commits himself to do likewise. What makes Billy's speech act special is that it directs Charley's *linguistic*, not his *physical* behavior (as in "Move the desk") and commits himself as well to perform linguistically (not physically). In effect, what Billy is doing (assuming that Charley tacitly agrees to go along) is setting the table for what is going to happen linguistically (what the text will be about in the immediate future) and, insofar as he is doing that, he is changing the context of their discussion (their speech activity or discourse) more radically than he would have had they continued talking about Jeter.

As organizers of our language use, and as context-changers or context-initiators, master speech acts do much work. They are everywhere. When a news commentator titles his text "China revisited" his master speech act commits him to focusing his attention on China (not Korea or football). When a writer labels her work "A novel," we expect some non-historical accounts of what happened between 1914 and 1918—the years covered by her novel. When a minister of the church tells us "Let us pray" we know not to interrupt him with comments about Derek Jeter; but instead to expect speech acts that express thanks to God, and so on. When Daddy says to his four-year old daughter "Once upon a time" we, again, know what sort of speech activity (i.e., a series of speech acts) will follow. When a judge says to a witness "Well, what happened?" she expects a report, not an evaluation, about what happened.

Some texts need not be announced that way. Charley may start a conversation with Billy by saying "He did it!," leaving Billy to listen until he figures out that "he" refers to their friend David, and that David has gone to the marine recruiting station and signed up. There are also settings when there is actually a need for a master speech act, but none is offered. Here we naturally think of the physics professor who, mindlessly, starts his lecture without saying anything like "Today our topic is black holes." In all these and similar settings, we can maybe figure out what is going on, but in most of them, it would have been nicer had the speaker issued a master speech act to help orient us a bit sooner.

The ordering that master speech acts bring about is enhanced by what might be called speech act modifiers.[3] These modifiers are like master speech acts, but they play a secondary role only. After Billy and Charley have talked extensively about Derek Jeter's last year's on-the-field accomplishments, Billy might ask "What kind of a teammate was Jeter that year? Was he a leader, or what?" These questions are directives addressed to Charley that still fall under the aegis of the original master speech act. They modify, but do not completely change, the original one that helped start their discussion. What modifiers do is take their discussion in a somewhat different direction. Billy's modifiers also change the context somewhat, but only somewhat. Like the ruling master speech act that started the Jeter discussion, Billy's modifiers play the role of context initiators.

Stoppers[4] also have a role to play in this discussion of master speech acts and the modifiers associated with them. Obvious examples of stoppers are "Amen" that ends a prayer, "And they lived happily ever after" that ends a child's story, and "Let's talk about something else." Also, a series of connected speech acts can be stopped by the hearer simply not saying anything or walking away. Stoppers not only end certain kinds of speech activity, they also tell us that the general context we have been operating under no longer applies. With the end of a prayer, we are now allowed, if we wish, to discuss the on-going football game between the New York Giants and the San Francisco Forty Niners. This means new speech activity and new contexts.

At first blush, it appears that master speech acts and their associated concepts are not part of the text. The announcement of a story is not, after all, part of the story. Yet, master speech acts themselves generate their own context, and so seem to be acting as if they are part of the text. Just like any other speech act, one doesn't understand such an act unless he or she has some understanding of the context it activates. Thus, the master speech act, "Tell me more about the features electrons possess," can't be fully understood by either the speaker or hearer if neither has at least a crude understanding of atomic theory in particular and physics more generally.

More needs to be said about master speech acts to help us understand the roles they play in language. Often, even if not always, these acts direct future linguistic behavior by indicating that the speech activity (discourse) that follows has a certain kind of content. But master speech acts can also direct the point or purpose of this activity.[5] We expect that a large number of assertives (i.e., claims assessable as true or false) will follow when a scientist lectures us on the topic of electrons. But when the speaker is a preacher, we expect her to issue a wide variety of directives. We, she tells us, are supposed to do this and refrain from doing that. In contrast, a politician will issue many assertives, but will favor commissives. She promises to do this for her followers, and then promises to do that as well. She will solve all the peoples' political problems. In contrast, teachers will issue more than their fair share of evaluatives as they give grades to their students.⁶

MASTER SPEECH ACTS STRENGTHS AND WEAKNESSES

Seeing that master speech acts help us to talk about almost everything, we might think there must be an infinite number of speech acts (texts) in the language we use; that, in turn, might tempt us to think that although our language appears to be working well, it really can't be because of its complexity. This skeptical thinking might lead us to suppose, finally, that concepts like master speech acts are not strong enough to restore order to our infinite language enterprises.

A more careful look at how master speech acts work helps, a bit at least, to dispel this skepticism. These acts do not bring order to the language all by themselves. When a judge tells a witness to do nothing else but describe what he saw at the scene of the crime, the witness's testimony is restricted to using language descriptively (i.e., using assertives). The witness is not allowed to make judgments of guilt or innocence. But the judge's master speech act does not give meaning to his request to describe what happened. The concept of describing is already built into our language. Also built in are certain grammatical structures concerned with the subject and predicate, with the past, present and future, etc.⁷ The master speech acts merely trigger usage that we call descriptive, directive, etc. Other such acts trigger other kinds of discourse already in place in our language. Our language has many devices to help bring order to what otherwise would be chaos. Thus, the infinite number of individual speech acts (and/or speech activity) that can be cited are, as it were, a part of an organization. Like an army of millions ordered into divisions, regiments, etc., our infinite speech acts are placed, at least in most cases, in certain units, or domains as I will call them, of thinking and so give order to our language.

How we identify these domains varies from thinker (philosopher and/or linguist) to thinker. The units could be identified in a way that matches, more or less, Searle's basic speech acts. In addition to descriptive speech activities, there would be speech activities that would be mainly composed of directives. But "mainly" is an important word here simply because even if a certain stretch of speech activity's main purpose is to direct the behavior of others, it would be difficult to avoid also issuing a host of other kinds of speech acts such as assertives and commissives. Hence, order needs to be invoked in terms of speech activity content and in other ways. The "language games" played could include scientific thinking and this form of thinking is commonly broken down into "hard" vs. social science thinking. Other games identified could include ethics, prudence, fiction, history, aesthetics, real games (such as baseball), informal talk over lunch (may not invoke master speech acts at all), religious services, religious prayer, etc. For the present purposes, whether this or some other organizational system is accepted does not matter much. What matters is that some sort of order needs to be (and is) already installed in our language.

SURFACE CONTEXTS

One of the many problems involved in discussing what I will call surface context is its rapid change. In a string of speech acts (speech activity, discourse) the context changes with each speech act on the string. As we will see, the change may not be great, but it is present. The context of the fifth speech act in a string, "I hit him hard," is different from the context of the next speech act, "He hit the ground head-first real hard." The "hit-him-hard" speech act and its context are presupposed by later speech acts. These rapid changes from speech act to speech act make one wonder, once again, how confusion can be avoided in understanding what is going on linguistically.

We might seem to be inviting confusion not just when one speech act is followed by another, but when one speech act (token) spoken in one context is followed by an identical speech act in a different context. A friend says to Sam "You should return the book," one that was borrowed from its owner; but then later says the same thing even though the friend realizes now that the book "owner" stole it from the book store.

Still another change can add to our sense of confusion as to what is going on linguistically. A speech act that at one moment is part of the context can, in the next, be part of the text. Consider Eloise who has forgotten that she borrowed a book from Eleni. But Eleni has not forgotten. When Eleni reminds Eloise about the book, her talking about borrowing moves into the text. It becomes part of their current conversation. But, of course, the next day, the matter of borrowing moves back into the context as Eloise returns the book to Eleni.

This last example makes it clear that surface contextual materials do not differ from textual ones just because of their force or content changes. Also, they differ because of where they stand in the layers of speech acts that form the context.

Why do surface context materials stand where they stand? Well, of course, because they are not part of the text, and because most of them have been uttered recently. Many (most?) of the surface context "speech acts" are factual in nature. They are there because both the speaker and the hearer(s) assume that there is no need to repeat what both parties know. Repetition, especially of what has been said recently, is time consuming and so slows the communication. It's like counting money hidden under the pillow that has already been counted. All one needs is to count the money gained today to know how much one has.

Of course, one may forget how much money is under the pillow. Then there must be a recounting. The easy-access feature of the surface context allows for such easy recounting. We say that the person who has forgotten speaks "out of context." She now speaks "in context" when she is quickly reminded what it is she has forgotten.

Speaking out of context happens in other ways. One of the participants to a discussion might have arrived late and is not privileged to all that has already been said. That sort of out-of-context talk can be corrected easily, again because the context can be accessed so easily. We take a few moments and get our tardy arrival "up to speed" in our conversation.

It is somewhat more difficult to correct another out-of-context problem since it involves engaging in research. An advertising agency may have, for example, deliberately spoken out of context by ripping away part of the text, thereby misleading its hearers. Copi in his introductory logic text gives such an example:

> Quoting out of context is sometimes done with deliberate craftiness. In the presidential election campaign of 1996, the Democratic vice-presidential candidate, Al Gore, was quoted by a Republican press aide as having said that "there is no proven link between smoking and lung cancer." Those were indeed Mr. Gore's exact words, uttered during a television interview in 1992. But they were only part of a sentence. In that interview, Mr. Gore's full statement was that some tobacco company scientists *"will claim with a straight face that* there is no proven link between smoking and lung cancer.... But the weight of the evidence accepted by the overwhelming preponderance of scientists is, yes, smoking does cause lung cancer.[8]

So far, in this first look at surface contexts, several of the points made already can be quickly restructured and restated as follows. Speech acts are not just made-up of linguistic forms. A sentence, even if grammatically correct, is not a speech act. Rather, such an act is an utterance by a speaker and addressed to one or more hearers.[9] It is this full sense of a speech act to which contexts attach. The attachment is close enough to allow what is surface context to move back and forth to the level of the text rather easily.

The closeness of the context to the text is manifested in other ways. As part of the full sense of what a speech act is, the speaker can change the text by invoking master speech acts or just moving from one topic to another at will. In doing either, the speaker alters the context. The hearer at the beginning of a conversation can later become a speaker so he, too, can change the text and context. The speaker and hearer change the context in another way, only in this case not by choice. As participants in a conversation, for example, the context tells us that each of us is a rational person who can understand the local language, can understand arguments, and has personal peculiarities of one sort or another that affect his judgment.

The text, by contributing force and/or content to any speech act, can also make a contribution. If the lecture concerns the moon, certain presumptions will be in place. Some have to do with gravity and others with the moon's general location.

Already it should be clear that the changes found in the text and in the surface context can be of two basic types. The text will change as one speech act follows another. Even if the text is operating within a confining master speech act, each act in the string that makes speech activity will be different in some minor or major way. Correspondingly, the context will take its lead from the text and change as well. New speech acts follow old ones, and new contexts follow old ones in parallel.

The second kind of change concerns the status of both ordinary speech acts and master speech acts. When spoken, both can be thought of as being part of the text. But then, in time, as our attention turns to other (related) matters, the texts morph and become part of the context. However, they can regain textual status when we are reminded what we said back then. Then, once again, as we turn our attention to other matters, they can turn back to being part of the context.

So what is the answer to the question "Are master speech acts part of the text or part of the context?" They are both. In the beginning they are probably best thought of as part of the text (indeed that part that helps to create more text), but quickly, these acts become part of the context. All this is very confusing, especially since, so far, we have only an incomplete picture of what a text is. Is a text what we say at just this moment, and thus expressed as just a single speech act? Or is it a speech activity concept that encompasses a string a speech acts? How long can the string be and still be part of the text? I am afraid we will have to wait for answers to these questions. Answering another question about the text will also have to wait. Is the distinction between the text and context as sharp as it seems? Isn't context built right into the text in a fundamental way so that the text cannot be articulated unless some contextual material is present in it from the very beginning?

NOTES

1. Tyler Kepner, "In Enemy Territory, a Farewell Comes With a Warm Embrace," *The New York Times,* September 29, 2014, D1, D2. To read more about Jeter see "No.2 Makes a Perfect Ten," *The New York Times,* December 7, 2016. This article is about retiring Jeter's No. 2.

2. N. Fotion, "Master Speech Acts," *The Philosophical Quarterly*, Vol. 21, No. 84, July 1971, pp. 232–243. See also N. Fotion, "Speech Activity and Language Use," *Philosophia*, Vol. 8, No. 4, October 1979, 615–638.

3. Fotion, Ibid., 240.

4. Fotion, *op.cit.*, 240.

5. Searle, "Taxonomy," 2–3.

6. N. Fotion and D. Seanor, "Basic Acts: Is Five the Magic Number?" *Philosophical Inquiry*, Vol. XXV, Winter–Spring 2003, No. 1–2. We add evaluatives to the five basic acts presented by John Searle. These acts rank almost anything on a scale appropriate to whatever is being assessed. Students, apples, cars, women at a beauty contest, weapons, computers, can all be ranked.

7. These concepts quickly lead to a discussion of the categories, and at this point we are reminded of Aristotle, Kant. Husserl, and more contemporary writers such as Gilbert Ryle, Roderick Chisholm, and Reinhardt Grossman. For the purposes of this study, it is not necessary to enter into the battle as to whose set of categories is preferred. It is enough here to merely point to what really is obvious that there is at least one layer of presuppositions that takes category claims seriously. More will be said about categories in later chapters. For a clear and comprehensive account of the category battles (what the categories are and even whether it is possible to find a set of categories) see "Categories," *Stanford Encyclopedia of Philosophy*, latest revision 2014 on the Internet.

8. Irving M. Copi, Carl Cohen, and Kenneth McMahon, *Introduction to Logic*, 14th ed. (Boston: Prentice Hall, 2011), 143.

9. This as we will see later is not quite right. But for now, it will do.

TWO
More Surface Material

MEANING AND MEMORY

All the types of contextual considerations discussed in Chapter 1 apply when Susan says to Tina: "I got a job today as a vice president of personnel with XYZ." Some of these considerations have to do with Susan and her educational background; and with Tina who also works for XYZ. Tina knows what it is like to work at the local offices and that getting a job at that office involves undergoing many interviews.

In addition to these surface contextual considerations, context has much to do with meaning. Language users presuppose that they know the meaning of the concepts found in their speech acts. Like most others, Susan does not talk about these presuppositions. Instead, she talks about her job. But without having a sense of the words and phrases in her speech acts (text), we, who are listening to her, might think that she is talking in a foreign language we know little or nothing about. Or we might even think that she is not using language at all but, instead, is just babbling.

In what sense does Susan know that she is in conversation with Tina, and that Tina knows what she is talking about? Certainly, not in the sense that she has the ability to recite definitions of the terms used by Susan. Indeed, both may be very bad at forming non-ambiguous and precise definitions. Yet we say both Susan and Tina know what she is talking about. For instance, concerning the concept of "vice president" Susan and Tina know that the position in XYZ carries social power, but only a limited amount. They know this because they also know that there are several vice-presidents in their local (regional) office and one president with much more social power than Susan has just been given. They also know that there are directors of offices who rank beneath Susan. In short, although they can't offer a definition of Susan's new position as vice president, they know how to use and characterize it because, in part, they both know the context in which this concept occurs.

To be sure, Tina's characterization of "vice president" most likely is not exactly the same as Susan's. But if they are to communicate successfully, there must be a considerable amount of overlap or sharing present. Robert Stalnaker calls this overlap the common ground.[1] I prefer the term common memory because it is more descriptive than common ground. In our case, the common memory "vice president" is being used locally. Only Susan and Tina are sharing. All others, who do not hear what they are saying to one another, are not operating in the same specific context as our two co-workers. This is so, even as is likely, that the others know how to use and characterize the concept vice president much the same way as Susan and Tina do.

MEMORY BANK

At this point, Susan-like and Tina-like stories create a sense of urgency to turn the dichotomy between text and context taken from Chapter 1 into a trichotomy. I will use the metaphor "memory bank" as the name for the third party to our system for understanding "context." It will turn out to be a very important concept, in some ways more important than "context" itself. Below are accounts, initial ones at best, of the bank and its parts. In these accounts, I move from the general to the specific.

Memory Bank. The "place" that contains all that one remembers however one remembers it. It represents all of that which is recorded in the brain and other parts of the body. Some of that might be tapped to do contextual work, but some might not be. The bank contains memories that are mostly shared with others, but also some that are not.

Three types of memory comprise the Memory Bank: Common Memory, General Memory, and Local Memory.

Common Memory. The coverage with these presumptions is universal or almost so. We all assume that any baby left at home in its crib last night is the same one in the morning. Normally, we don't even think of the possibility that a different baby will be there due to some radical, and totally unpredictable physical or social change. Nor do we consider the possibility that it will have transformed into a cute puppy overnight. More than that, we assume that the baby is real. It is not considered a figment of our imagination. It is not seen as a cluster of sense data, or any construction of that sort. Also, we assume that the law of non-contradiction holds when we talk about the baby. Still, for all its vastness, the common memory represents only a portion of the memory bank itself. It doesn't contain, for instance, idiosyncratic memories (e.g., what I read in my room when I was alone last Tuesday and what I whispered to my love last night at dinner).

General Memory. Coverage here is shared by ethnic groups, nations, members of various large social institutions (e.g., the Catholic Church, Google, major league baseball, various professional associations). Susan and Tina's understanding of vice president is an example of this kind of memory in the sense that a wide variety of institutions employ a similar concept and an even larger group of people have a rough sense of its meaning.

Local Memory. Here the memory of events extends only to smaller groups such as the members of a department at a meeting, extended family members gathering during holidays, attendees at a church dinner, a group of physicists at a lecture, and the like. In these events, the participants probably employ shared expressions that are found beyond their group, but only they have the memory of what was actually said during their time together.

As suggested already, this three-fold division of the memory bank is rough-and-ready on purpose. There is no reason to draw a sharp line dividing general from local memory. As we will see, it is enough to simply be aware that this and other distinctions mark significantly different memories with different characteristics. Further, it is useful to make a distinction now between what I call the actual range of what is remembered and the abstract range. The Susan and Tina story above is an example of actual range. Only they know about *their* particular use of the abstract term "vice president." The abstract range has to do with shared memories.

Memories are of course individual. We all carry our own memory storage. But we share memories with others for a variety of reasons. Some have to do with biology. Biologically, we differ from one another in various ways, but we are also alike in many ways. It is not surprising, then, that we have "like" experiences that we share with others. We also live in close proximity with many other humans. Their experiences, understandably, are like ours. So again sharing makes sense.

Separate from all the memories in the bank are memories found in an actual or real context. The items in the context memory are those that are actually used to support the text (speech activity). In effect, they are items that were sitting in the memory bank waiting to play a supportive role for a text in need of support, and now are being used. Once used, they return to the bank to be ready to be used again. Much more will be said about the context memory and the memory bank, but for now all we need to know in order to proceed is that these two concepts are in place sitting below the text ready to do their work. In summary, these three parts of our language use (and thinking) are arranged as follows.

1. The Text. (For sure, this is composed of individual speech acts, but, more likely, also of speech activity of various lengths.)
2. The Memory Context. (What shows up here supports the text so as to make it meaningful or make it more meaningful. What are in the context are items originating from the memory bank.)
3. The Memory Bank. (This category is located in the individual and constitutes all that the individual knows either in the form of knowing-that or knowing-how.)

The above discussion of the text, the context and the memory bank, presupposes that there are five major dimensions or types under which that discussion falls. The memory bank items feature one or more of these dimensions.

1. Range. This concept has just been explained. The main distinction to keep in mind is between abstract and actual memories.
2. Content. This concept has to do with the subject matter of a text or, to put it linguistically, to whatever we refer. It also has to do with the subject matter of the context memory and the memory bank itself. The content can refer to objects, processes, abstract entities, institutions, etc.
3. Major speech act types. This "dimension" applies to speech acts found embedded in speech activity. Some forms of speech activity will feature assertive (e.g., as in scientific discourse), others will feature directives and evaluatives (e.g., as in ethical discourse), etc.
4. Source. This "dimension" is concerned with how any particular item in the memory bank found itself in the memory bank. Was it produced by some speaker? Did nature put it there? We might suppose, for example, that nature forces us to react to gravity the way we do. Or should we suppose that a concept (such as a theory in the sciences) is produced by groups of specialists in the field? Now obviously, the source is human but it is not necessarily created by individual humans. Another possible source of an entry in the memory bank could be a psychological condition. I might react emotionally to going to school because a particular teacher treated me badly.
5. Stability. Some trivial fact might be put in some person's memory bank, be used in discussion soon after, and then completely forgotten. It could then be said that this trivial fact is unstable. We can't expect it to be around three days later after many other things have been discussed. Other entries in the bank are more stable because they are shared by a large group that cannot, as a group, completely forget the item. Still others can be thought of as unforgettable such as the effect gravity has on our behavior on the planet.

Of the five dimensions, content will turn out to be the most important, mainly because it controls the other dimensions more than they control it. The results of motivation studies in psychology, for example, cannot help but be relatively stable since they tend to be made public and so are shared in the memory banks of a large group of professionals. Because, then, content shows the most promise as a controlling dimension, the chapters following this one will canvas that dimension. Put differently, if the goal of this study is to highlight the complexity, variety, and flexibility of what most generally the study of context falls under, the place to go is to the memory bank since that is where all the content related to context is found.

GRAMMAR

In certain situations we can say "Shoot," "Duck," "Run," and "Watch out" and not appeal to grammar directly. But the overwhelming number of speech acts we issue requires such an appeal. To use our natural language we must come to know about the distinction between subject and object, past, present and future tenses, conditional and non-conditional speech acts, between verbs, adverbs and adjectives, etc. In our memory bank we know about these distinctions mainly as knowing how. We are normally not very good at explaining the difference between, say, adverbs and adjectives, but we (well, most of us) do know how to use these concepts and do so over and over again. What we know about these distinctions turns out to be stable and perhaps universal. These concepts don't change overnight, and they have to be shared by everyone in any one culture and even in clusters of cultures if we desire to communicate with one another.

What roles, one might wonder, does grammar play in the discussion of context? It might be thought none at all since context items seem to do their work on already formed speech acts while grammar is concerned with the formation of speech acts. It is true that most of the appeals to context so far cited in this study play roles in helping to make sense, or helping to make more sense of already formed speech acts, but that doesn't mean that these are the only roles context(s) can play. Both not-yet-formed speech acts need help to become formed and that help can only come from the context that gains its content from the memory bank concerned with grammar and meaning (see above). If that is right, context's role in language (and thought) can be either supportive of formed speech acts or creative of speech acts in the process of being formed.

NOTE

1. Robert Stalnaker, *Context* (Oxford: Oxford University Press, 2014). Stalnaker uses "common ground" which means "a body of information that is presumed to be shared by the parties to a discourse," 2.

THREE
Social Contexts

THREE VIGNETTES: MAINLY NON-NORMATIVE (ASSERTIVE) SOCIAL CONTEXTS

Consider the following setting. Abe says to Bruce: "Charley is a typical American." Bruce, it turns out, knows little of America so he responds by saying, "Why do you say that?" Abe might not find it easy to respond. He can easily identify someone as an American, even on a crowded street in Paris. But he finds it difficult to articulate what is in Charley's countenance and behavior that turns him into a typical American. Does it have to do with how Charley talks, walks, dresses? Is it his haircut? Or maybe it has something to do with his social, political, and religious views.

Abe eventually responds by appealing to his context memory (and before that to his memory bank). If Abe gave the following account, we would wonder if he knew what he was talking about. "An American is quiet, self-effacing, formal in manner and dress, thoughtful, and very much an intellectual." To get it right, Abe would have to give us a quite different list. You and I could help him out. You might list traits a, b, c, d, and e, while I might come up with traits a, b, d, e, g, and h. That each list diverges a bit from the other is not a problem. The content of concepts, such as typical American, German, Catholic, Republican, are inherently vague and ambiguous. But they are not as vague and ambiguous to encourage us to accept a list of shared traits that do not include any of those traits in the two lists. The content of these social concepts may not be very clear, but they are clear enough. And what they tell us is that many (perhaps most) Americans have most of the traits we have identified as part of what we mean to be a typical American. Putting it differently, each of us has a loosely shared memory of what an American is like.

Consider another vignette. David says to Edward: "By and large, Catholics are social conservatives." His is a local example like above. Both are local in the sense that Abe and Bruce, in one case, and David and Edward, in the other, are the only ones participating in these discussions. Only they have memories of whatever is said. What is local about them has to do with a series of speech acts (i.e., speech activity) found in the text itself. In the abstract sense, their use of "Catholic" is general since it is a term used in a sharing way with a wide variety of people.

In their discussions, if Abe and Edward refer disparagingly to Catholics as "Licks" and if they are careful never to use this term when talking to others, their use would be local. But still, "Licks" could have the status as a general term if it were known and used by large groups of citizens.

There is an indefinite number of terms like "Catholic" that can be used descriptively, although they can be used evaluatively as well. They infuse our (shared) memory bank. Present here also are beliefs used mainly descriptively such as "They (the XYZ people) have a high crime rate," or "Their women act and dress very conservatively." Typically, these social concepts and claims have the following features. They vary as to their generality. Some, like "Catholic," are used inter-culturally while others like football (American style) are mainly intra-cultural. Some terms and claims are even more restrictive.

Another feature of these terms and claims is that they tend to be stable. They can change as, perhaps, the concept of marriage is changing at present in the USA. But especially those concepts and claims associated with a formally structured institution (e.g., Chamber of Commerce) resist change.

Still another feature of these kinds of social claims is that they are correctable. They are, after all, empirical claims and with such claims things can go wrong. My memory bank might tell me that Jews, on average, are no better educated than the rest of the society, but I might be wrong about that. My views might change if some sociologist showed me reliable data proving otherwise.

In Abe's characterization of a typical American, the content of his speech is also empirical. He is being asked to report how "typical American" is used. However, in his response, he badly misreports what the shared meaning of the term is. Although Abe can be corrected in his usage, the correction does not necessarily involve changing the meaning of "typical American" and thus encouraging its demise. The actual range of his misreporting on the text level is local since he is talking to Bruce only. It is similarly local on the context level since, again, only Abe and Bruce are aware of their conversation and the context that goes with it. On the abstract memory level, the range is general. "Typical American" is an expression known by many Americans and non-Americans, but not by everybody.

NORMATIVITY

Consider now some quite different vignettes. These come from Barbara Tuchman's *The Guns of August*. Her book is about August 1914 and the before and after events related to the start of World War I. She describes in the following passage the thoughts about war of a British poet and a German writer.[1] Tuchman writes as follows:

> People entered the war with varying sentiments and sets of ideas. Among the belligerents some, pacifists and socialists, opposed the war in their hearts; some, like Rupert Brooke, welcomed it. "Now God be thanked who has matched us with His hour," wrote Brooke, conscious of no blasphemy, in the poem "1914." To him it seemed a time:
> To turn, as swimmers into cleanness leaping
> Glad from a world grown old and cold and weary...
> Honour has come back...
> And Nobleness walks in our ways again,
> And we have come into our heritage.

Some Germans had similar emotions. Thomas Mann wrote that war was to be "a purification, a liberation, an enormous hope. The victory of Germany will be a victory of soul over numbers. The German soul," he explained, "is opposed to the pacifist ideal of civilization for is not peace an element of civil corruption?"[2]

With these passages we are immersed in normativity. Brooke and Mann praise war, while Tuchman does not. In addition, Brooke may be praising British culture simply because a portion of it appreciates the purifying effects of war. Mann is more direct in praising German culture. However, if we allow both writers their judgments, they need to give us an account of why Britain and Germany need to be praised. With such an account or justification, it may be impossible to avoid appealing to facts and other norms than those found in their texts. Mann, in fact, gives reasons in other passages for claiming that Germany has a superior culture. Germany, he says, does better than a decadent France and England that stand for "selfishness and perfidy." Germany, in contrast, "being the most educated, law abiding, peace-loving of all peoples, deserved to be the most powerful, to dominate, to establish a German peace."[3]

Expressed in terms of the concept of context, what has Mann done by giving us his account of Germany's cultural-superior status? Well, perhaps nothing by way of appealing to his memory bank. If his thoughts and arguments are original with him, then he has little or no need to appeal to the context *on the social level* of his memory. He merely states what he wants to say *on the text level* and that is the end of the matter. No doubt he cannot avoid appealing to the context on some other level. However, it is unlikely that his thoughts (both descriptive and normative) are totally original with him. To the extent that they are not Mann's, he would be bringing into his text materials from his memory bank to his context (memory). It is as if he were moving a piece of furniture from the basement to the main dining room.

Some additional comments on Mann are in order. Clearly his text is public in the sense that anyone who wishes to access his content can do so. Further, anyone can agree or disagree with him. Clearly, also, Mann's text is immersed in a rich context. What he has done is take his memory material, some (much?) of which he shares with others and put it to work as context for his text. He even raises some of his context material to the level of the text while he leaves the material in his memory bank in its place. It is as if, to use another metaphor, he has taken a Xerox copy of a document with him to the context level, and possibly even to the text level, while leaving the original where he found it.

Finally, it should be clear that in starting and finishing his analysis of Germany's social status, he is operating on the local or, possibly, the general level. The context to which he is appealing has largely to do with Germany. His context tells us about the (wonderful) characteristics of that nation's people and is designed to engage Germans especially. So, anyone who wishes to understand and appreciate his text has to be familiar with entries in their German context memories.

But notice that non-Germans who are familiar with Germany will not likely appeal to the same memory bank. British Empire individuals will have their own memory banks concerning empire, democracy, competition among the major political parties, etc. As a result, the Germans and British will see the world in different lights as they are reading Mann's text. They, we might say, have different memory profiles. There will be different lights even within Germany. If a German communist reaches into his own memory bank, he will pull out different ways of viewing the onset of World War I and (strongly) disagree with Mann. Thus, the text and the context that Mann brings into the text is not a source of a definitive answer favoring the communists, the royalists, or some other view of politics. All sorts of political positions have a place in the memories of the German (and British) people. They will be in place in the memory of each person who is informed about the political options of the day. Still, it is surprising that even though the context memories of various groups in our examples differ significantly, most of those who read Brooke and Mann can understand what they are saying. Presumably these readers do so because there is still a significant amount of sharing present.

APOLOGY

It is with apologies that I present these vignettes and later, both more vignettes and cases, as if they give us the whole story about context. I engage in this slice-by-slice analysis for expository reasons only. One needs to look at each level of thinking and speaking on its own to see what features it possesses. Once that is done, one can think holistically again and thereby gain a view of the big picture.

NOTES

1. Barbara W. Tuchman, *The Guns of August* (New York: The Macmillan Company, 1962), 311.
2. Ibid.
3. Ibid.

FOUR
Ethical (Moral) Memories

THE COMMON MORALITY

There are certain ethical principles, no matter how they are expressed, that are held by the vast majority of people living in the vast majority of cultures on this planet.[1] These principles are so commonly held that most theoreticians in ethics are anxious to show that their own high-order principles do not compete with them. Few, if any, want to say that their theory does not generate, produce, or exhibit principles such as "Don't harm others," "Help others in distress," "Do not deceive others by lying, withholding information, exaggerating, and the like," and "Act justly." These moral or ethical principles can be alluded to or actually cited. If cited, and thus raising them to the text level, the speaker can say something like "You ought to tell the truth to your wife about your dalliances because it is wrong to deceive."

This sort of appeal to the context represents the simplest form of justification found in ethics. On the text level, a normative (prescriptive) claim is issued ("Tell your wife . . ."). In this way, the issuance is supported by a rule or principle found in the memory bank, and then, if needed, is raised to the level of context. It can, as in this case, be raised all the way to the text itself. More often than not, however, not even this form of explicit reasoning is needed. The speaker simply tells us what he ought to do (the text) and lets the context do its work silently. The assumption in these settings is that the principle in the context is so well known, that is, shared by almost everyone, that there is no need to cite it. We cite the principle or rule only when our listener is stupid, thoughtless, or a little bit (or very) bad. The no harm (e.g., "Don't kill") rules work the same way. Almost every day we don't harm hundreds of people with whom we come into contact; we don't give it a thought. Most of the time we act from habit when the occasion for doing so warrants. These habits are strong. But they are not above being "broken." Criminals know this all too well. But the strength of these habits is such that even these bad people don't harm the vast majority of those with whom they come into contact. They only harm those who work in jewelry stores, banks, etc.

NOT QUITE SO COMMON PRINCIPLES

There is a group of other rules and principles that have almost the same status as the common morality. Many of them overlap the common morality to some extent. But what usually marks them as different is that they are somewhat more specific. Here is a sample list.

1. You ought to pay your debts (and do so promptly).
2. You should be grateful when others help you.
3. You should be polite and friendly when dealing with others.
4. Be thoughtful.
5. Be diligent when you do good work.

These memory entrants have a very general range but not quite as general as the principles in the common morality. But they too reach the status of habits for most of us.

There is another set of principles sitting on a higher (or lower) level. These are the kind loved by philosophers and religious leaders. More or less what these principles do is abstractly blanket all or most of what we call ethics (morality). Any one of them can be stated as the one major principle or as a small cluster. Here is a sample list of these super-principles.

1. Never treat others merely as means.
2. Maximize good consequences and minimize bad ones.
3. Act in accordance with God's will.
4. Strive for the virtuous life.
5. Strive to live in ways that your actions will reflect rules and principles that cannot be rejected by others.[2]
6. Love your enemies as you do your friends.

Two points should be clear about these super-principles. If, as I claim, they are found someplace on the ethical level in most, or some, of our memory banks, there seems to be little in that portion of our banks that tells us whether we should prefer one of these principles over the other. To be sure, there is pressure to have a preference here since these principles are taken by many thinkers to be in competition with one another[3] and, often, we have been educated by our religion and/or schools to prefer one over the other. In this sense, these principles are quite unlike the common morality. Each of the common morality rules or principles (e.g., "Don't harm others") stakes out a portion of life different from the portion staked out by the other rules or principle. So, these common morality principles don't usually compete with one another in the same way except on rare occasions. But each of the super-principles is said by some to be catholic and so seemingly we cannot avoid coming into conflict. The claim here is that, sitting where they do in the ethics vault of our memory banks, none of these principles can gain traction to "defeat" the other.[4] Defeat of all but one of them may be possible but again the claim is, in order to bring about defeat, an appeal must be made to one or more of the deeper levels of the memory bank (e.g., where we find logical principles).

For now, however, all I am doing is making a claim about conflict resolution or lack thereof when one is operating on the super-theory level. Justifying this claim must come later after the discussion moves to consider the roles that still other levels of the context memory play in our thinking (and language use).

Second, some of these super-principles have, in part, a meta-ethical rather than an ethical status. As such, they do not directly guide us in our lives as does "Tell the truth." Rather, they help us figure out how we should be guided.

Third, some of these super principles have a rather limited actual or abstract range (item 5 above). Only a few philosophers and other scholars know about and understand them. Others on the list are understood more generally (item 6).

An additional point needs to be made about all the rules and principles discussed above, not just the abstract ones. Many of them are stated in such a way that they sound absolutist in nature as if when we say "Don't lie" we mean "Don't ever lie." But most of us know that this rule and even the rule about not harming others are not really absolutistic. Most rules have exceptions. But on the memory bank level so far discussed in this chapter, there is, at best, only a few tools, procedures, etc. that help us to know when we should make exceptions and when we should not. How we can make exceptions has to be made clear with the help of one or more of the other levels of our memory banks. The rules and principles of the common morality, for example, do not speak to how they can "undermine" themselves. All these rules and principles tell us what the general direction of our behavior should be or what general virtues we should pursue and develop.

These points about exceptions can be stated in another way. So far, I have assumed that entries in our memories are individual speech *acts*. It is as if arguments and theories composed as they are of speech *activity* (*discourse*) don't belong in our memories. It is also as if these arguments and theories are found mainly on the textual level—not on any of the levels of our memory banks.

For now, let us assume that this is correct. After all, developing and coming to understand an argument or theory requires conscious thought that puts individual premises (speech acts) together. More than that, the premises must be put together logically. And that means that an appeal will have to be made on the level of our memories beyond that of the ethical. Again, the ethical level itself does not tell us much of anything about logic. Its main task is to do something else, that is, to provide us with normative premises.

But even adding logical considerations to the mix is not enough to help us put together a meaningful normative argument or normative theory. Additional premises, empirical ones, are most likely required. We may be dealing with very simple arguments of the kind "Don't steal because doing so goes against the rule that forbids stealing" that do not appeal to empirical premises. But the more complicated ones do, as we will see in the discussion in the next chapter. So, the pattern for normative arguments is to have them stated on the text level but have them well stated only when they borrow: (1) normative speech acts from our normative memories; (2) empirical speech acts from our empirical memories, and (3) logical principles from other parts of our memories. Arguments and theories are composites of speech acts and ways of thinking coming from different sources. There is no such thing (except in the simplest cases) as a moral argument or theory created exclusively from the material found in our normative memories.

MID-LEVEL THEORIZING

There is still another level of thinking, theorizing, and speculating in ethics (morality) that needs to be considered. That level is sometimes called mid-level theorizing. It seems to fall between a list of common morality principles discussed above and the super (and/or meta)-principles. The best-known examples of such thinking are Just War Theory[5] and various versions of Stakeholder Theory in business.[6] These theories are often justified by an appeal to some super-theory or even some combination of super-theories. Still, they seem to develop a status of their own in part because, interestingly enough, they can be justified by more than one of the super-theories. Thus, one can hold to Just War Theory and do so for Kantian-like reasons,[7] utilitarian reasons,[8] rights-based theory,[9] and so on.

Given that there are at least many levels of thinking in ethics, one might expect that the rules, principles, and even individual prescriptions would possess varied characteristics. As we have seen, common morality principles have an (almost) universal range, but it is not obvious that many of the super-principles do. Putting it differently, the common morality principles are in (almost) everyone's memory, the super-principles are not. As to the mid-level theories, they by definition have a limited range. They cover only some aspects of our lives (e.g., war only) and so ranges over fewer people. Their range of application is general, at best.

NOTES

1. Tom L. Beauchamp, "A Defense of the Common Morality," *Kennedy Institute of Ethics Journal*, Vol. 13, No. 3, 2003, 259–274. See also W. D. Ross, *The Right and the Good* (Oxford: Clarendon Press, 1930). Ross has a group of what he calls moral intuitions similar to Beauchamp's list of principles ("standards of action").

2. T. M. Scanlon, *What We Owe to Each Other* (Cambridge, MA: The Belknap Press of Harvard University Press, 1998). See Introduction.

3. Derek Parfit, *On What Matters* (Oxford: Oxford University Press, 2011), 410. Parfit holds the opposite view. For him, the competition between the major theories is only apparent since what differences there are between the theories will be resolved once we all reach the top of the magical philosophic mountain.

4. Nick Fotion, *Theory vs. Anti-theory in Ethics: A Misconceived Conflict* (New York, Oxford: Oxford University Press, 2014).

5. Michael Walzer, *Just and Unjust Wars*, 4th ed. (New York: Basic Books, 2006). Among scores of books on this topic see the following as a more recent example: Steven P. Lee, *Ethics and War: An Introduction* (Cambridge, New York, etc.: Cambridge University Press, 2012).

6. R. Edward Freeman, *Strategic Management: A Stakeholder Approach* (Boston: Pitman, 1984).

7. Stephen Coleman, *Military Ethics: An Introduction with Case Studies* (New York and Oxford: Oxford University Press, 2013), 18–24.

8. William H. Shaw, *Utilitarianism and the Ethics of War* (Abingdon Oxon OX: Routledge, 2016).

9. Kai Draper, *War and Individual Rights* (New York: Oxford University Press, 2016).

FIVE
Empirical Contexts

OVERVIEW

There is a parallel between our moral memories and our empirical ones. In the former realm there is the common morality, in the latter there is what I will call common (or basic) beliefs. These basic beliefs or memories are numerous and varied. They include the following: real people are out there in the world, we see and interact with them. Mountains are there, too, as are trees, bushes, and grasses. Objects fall to the ground when released from our hands. The ground is generally hard, so animals can walk and run upon it. There is a lot of water in the oceans, lakes and rivers. Flaming sticks burn our hands if we touch them. Humans don't live forever. We humans have feelings and emotions. I could go further. These are empirical or empirical-like things we know now and knew about well before modern science started changing our lives.

In addition to the parallel between our moral memories and our empirical memories, there is another large class of memories that were clearly in place before modern science got going. In one sense, these are not common memories that we share with everybody. But they are common in the sense that we all go through these same kinds of experiences. Everyone has memories of a conversation with a friend, going to the bathroom in the middle of the night, eating in for dinner, getting a stomach ache, almost falling when he stumbled upon a rock, on and on.

Together, both kinds of experiences, the truly common ones, and the ones like those everyone has but nonetheless are idiosyncratic, dominate our memory bank. Even scientists in specialized fields spend an enormous amount of energy each day doing and maybe remembering these kinds of ordinary things. Most of what happens in their lives is ordinary rather than extraordinary.

Life, and the special memories that are there for scientists, are even more ordinary for non-scientists. Ordinary people use cell phones, computers, and TV equipment, drive cars, and fly on airplanes, but their memories are not likely infused with information about how these wonders of the modern world work. They just use these wonders and talk about them in ordinary ways.

The abstract range of the memories (of discovered facts, laws, principles, and theories) in science is much smaller. A physics lecture on black holes is not likely to draw large crowds. Some other fields, perhaps biology, might draw a bigger audience. But even a lecture in psychology done in the spirit of Freud will not likely draw crowds on the level of a National Football League game. Sharing in science is on the general level rather than the common level. At times it may approach the local level as when scientists work on a very specific topic. It could be that only a few hundred people around the world, and the eight or ten at a particular meeting on this topic, know what is being talking about.

So although we may suppose that science dominates our lives, it doesn't dominate our memory banks. The empirical part of our memory bank is full of ordinary memories that are ready to support our ordinary texts by raising certain of these ordinary memories to the context level. The empirical part engages in the same process of raising memories to the context level for the more scientific portions of our lives.

There are some differences. Setting aside the basic beliefs (e.g., it rains) which are pure facts or laws, texts created by ordinary people are not likely to be pure. They are made up of a crazy mixture of empirical claims (assertives) and normative claims (directives, commissives, evaluatives), expressives, and even occasionally declarations. Scientific texts, not surprisingly, are more disciplined. There are times in a scientific text where only (or mainly) assertives are issued. Also, master speech acts are used to control the direction and nature of the text. In addition, logic will also be appealed to in order to maintain order. More on that in chapters to follow.

NETWORK AND BACKGROUND

At this point, it is useful to introduce a distinction that John Searle makes in connection with his views on context. The distinction helps to show that there is more to empirical contexts and memories than one might suppose. There is, Searle says, a difference between what happens in the Network and the Background.[1] In his book *Intentionality*, he talks about these matters in mental, rather than linguistic terms. But much of what he says about the mind applies to language.

> Intentional states with a direction of fit have contents which determine their conditions of satisfaction. But they do not function in an indepen-

dent or atomistic fashion, for each Intentional state has its content and determines its conditions of satisfaction only in relation to numerous other Intentional states.[2]

Searle here is talking about what he calls the Network. Translating his message, the Network covers pretty much all of the speech acts and speech activity discussed in this and previous chapters. For Searle, intentional states (speech acts and speech activity) belong to his (holistic) Network if they can be naturally thought or talked about in linguistic terms. But, for him, not everything can be naturally talked about in these ways

> I believe that anyone who tries seriously to follow out the threads in the Network will eventually reach a bedrock of mental capacities that do not themselves consist in Intentional states (representations), but nonetheless form the preconditions for the functioning of Intentional states. The Background is "preintentional" in the sense that though not a form or forms of Intentionality, it is nonetheless a precondition or a set of preconditions of Intentionality.[3]

Thus, there is still more to each person's context (and his/her memory) than we might suppose. Indeed, this something more has already been discussed briefly in earlier chapters and will need to be discussed in more detail in this and later chapters. But for now, in a chapter concerned with matters related to the empirical domain, only empirical domain Background factors will be discussed. In this regard, Searle is obliging in providing us several examples.

> Think of what is necessary, what must be the case, in order that I can now form the intention to go to the refrigerator and get a bottle of cold beer to drink. The biological and cultural resources that I must bring to bear on this task, even to form the intention to perform the task, are (considered in certain light) truly staggering. But without these resources I could not form the intention at all: standing, walking, opening and closing doors, manipulating bottles, glass, refrigerators, opening, pouring and drinking. The activation of these capacities would normally involve presentations and representations, e.g., I have to see the door in order to open the door, but the ability to recognize the door and the ability to open the door are not themselves further representations. It is such nonrepresentational capacities that constitute the Background.[4]

It is not that the Background cannot be represented in the mind or in language. It can, of course, as Searle's own writing on the subject shows. Rather, the Background carries all the items that fit into our memories and our contexts as dispositions. We know about the hardness of the floor primarily by knowing how to deal with it. Its hardness enables us to stand, walk, or run on it, causes us fear if we fall, enables us to place things such as chairs and tables on it, and so on. Similarly, we know about doors when we learn how to open and close them, how to hang things from them, and how to clean them. And because we know about these things dispositionally, we know about them without even thinking or talking about them. Again, we can talk and think about them, if we wish to.

In addition to making the Network/Background distinction, Searle makes another useful distinction.

> A minimal geography of the Background would include at least the following: we need to distinguish what we might call the "deep Background," which would include at least all of those Background capacities that are common to all normal human beings in virtue of the biological makeup—capacities such as walking, eating, grasping, perceiving, recognizing, and the pre-intentional stance that takes account of the solidity of things, and what we might call the "local Background" or "local cultural practices," which would include such things as opening doors, drinking beer from bottles, and the pre-intentional stance that we take toward such things as cars, refrigerators, money and cocktail parties.[5]

Thus, it should be clear that "the empirical" domain is not really a single domain found in one section of our memory bank. There are at least two basic "levels" or sub-domains. First, some empirical claims are accessible by every normal human being even those without any training in science. We all know how to identify trees, grass, rocks, hills, mountains, wind, and rain. We all also know about certain laws of nature, crude ones to be sure, such as that physical objects tend to fall to earth when released, that many trees lose their leaves as the weather cools, that humans rarely live to be 100 years old, that some objects such as diamonds are harder than most others. Roughly speaking, these are what I am calling common beliefs.

Second, there are those empirical claims found in the various sciences that are assessable mainly to those with special training. By and large, these specialists will know everything that we ordinary folk know about, but they know more. For instance, some specialists will know not only that objects tend to fall when released, they will know a formula ($S = \frac{1}{2} gt^2$) that enables them to know exactly how fast and how far an object has fallen. Others, with a medical background, will know not only that aspirin diminishes the effect of pain, but know with some degree of precision what counts as taking too little or too much of the stuff.

When first thinking about it, one might suppose that the specialist's knowledge would more likely be found in Searle's Background; and the generalist's more superficial and everyday knowledge would be found in the Network. But it is just the other way around. It is everybody's everyday knowledge that tends to take form as habits and so disappears from consciousness. Our contact with gravity when we walk, struggle to get out of bed, and jump when exercising is so pervasive that we cannot afford to think of these acts constantly. If we constantly attended to these acts, we would have little or no time to attend to our acts of work and pleasure.

To be sure, some activity when we engage in science also reaches down to the level of habit. Good scientists make observations in such a careful and objective manner so often that they do not have to give a thought to what they are doing. Instead of thinking about how they observe the world, their thoughts focus on observing the world. But it is another matter when they deal with formulas, both complicated and ones that are used by them only from time to time. With some exceptions, these formulas need to be left on the conscious level and so belong in Searle's Network. That means if scientists issue texts concerned with their research, these texts will be affected by the context primarily because they can call up the formulas from their memory bank and then place them in their context. Or, if this material is complicated, they might move it into the text itself.

Let's assume that a scientist has a strong memory and so successfully calls up the formulas on a regular basis. Yet, at times, his memory fails. What should he do? In science at least, he doesn't have to give up since he can access what can be called memory helpers. These helpers are other scientists, books (both textbooks and scholarly ones), articles, visual programs, computer-based programs, etc. In science, these helpers form the most organized sources of information available to our scientist to aid his failing memory.

Here is how the helper process works. The scientist accesses a book (or whatever) that is rich in materials about which he wishes to talk or think. The accessing process helps him re-remember. Now he can do his writing since the forgotten formula has been temporarily placed on textual level. Once his writing is finished, he can allow the formula to settle into his memory bank. If, a few months later, his memory fails him again, he can repeat the whole help process.

As described thus far, the process depends on the scientist remembering something. He needs to remember that he knew the formula at one time and, being a good scientist, he needs to remember where he needs to go to get the information. But what if he is not such a good scientist and so is completely baffled as to where to turn for help? Wherever he gets help, the issue will no longer be about memory. There is nothing to be recalled. The process aimed to get him to continue his writing will now be one of learning instead.

And that gives us a clue about how our not-so-well-informed scientist will (should) proceed. He will talk to his colleagues who, hopefully, will educate him about the books, journals, etc. that he should access and by talking to him about his academic failures. Once that is done, the process will be much the same as with the good scientist. The difference will be that when our deficient scientist comes to understand the formulas he never knew about, he will have added thoughts to his memory bank that were never there before. He will be enriching his memory bank.

MEMORY HELPERS

These variations as to how memory helpers work naturally raises the question about their status. Are they part of the context? More specifically, are they part of the memory bank? Or are they separate from it? To answer this question, it is useful to distinguish between several kinds of memory failure. It is also useful to make a distinction between the memory helper and what is remembered (the content of the memory).

1. Sometimes, simple memory loss can be remedied with no outside help. A name or a formula is forgotten. Instead of seeking help, the forgetful one racks her brain to try to come up with the memory. She fails. So she decides to put the strain of remembering aside and engage in some other activity. She knows, from past experience, that this tactic works at least some of the time. So she watches TV for a while and, surprisingly the lost memory pops up. This is a case of self-help where a portion of the memory bank helps find the lost memory content. The whole process (involving the helper and the content) is internal to the memory bank.
2. Another memory loss is remedied by an external reminder. John knows Sally's last name but can't remember it today. Nor, as it turns out, is he able to come up with the name on his own, try as he may. But he is talking to his friend Alice who, when asked, says "Her last name is Bickerstaff" and, just like that, John remembers. He even says, "Now I remember." In this sort of case the kind of help Alice gives is outside the memory bank but the content of what is remembered is inside.
3. Then there is memory loss of a complicated scientific formula (or theory). Here John needs to hear more than a recitation of the formula. Now Alice must both recite that formula and explain how it works. What Alice needs to do is help John by giving him a lecture since he never understood the formula even in graduate school. Here it is probably best to say that both the help given and the content of the formula are outside John's memory bank even though John had a vague memory of the formula's existence. A variation of this form of memory loss has John not talking to someone to get help, but directly accessing some textbook. In this case, both the help he receives and the precise nature of the formula are not in his memory bank. The only part that might be thought of as internal to his bank concerns his knowledge that accessing textbooks can be helpful. In another variation of this case, John understood the formula in his graduate school days, but now cannot

recall or understand it. Here the help is outside his memory bank, but it may not be clear what the status of the formula's content is in his bank. If, when he is reeducated about the formula his reaction is "I just don't remember (what I knew in the past)." Then probably it is best to say that both the help and the content of the formula are outside his memory bank. If, however, he (sincerely) says "Now I remember" then it is best to say that the help was outside his bank but the content is inside.

4. Then there is memory loss where not even hearing a lecture or reading a textbook will help John. His problem is Freudian-like. John evidently has some problems with his unconscious. He is suffering from a classic Freudian problem. He knows (unconsciously) that he hates his father but cannot admit that he has this hate consciously. The process here is somewhat like #1 and #2. The main difference is that as a helper Alice has been replaced by a psychiatrist. Freudian analysis would suggest that the memory always had a home in the bank, but John isn't even aware that the memory is there. Here the help is outside John's bank, but the content is inside.

NOTES

1. John Searle, *Intentionality: An Essay in the Philosophy of Mind*, Cambridge University Press (London, New York, etc.: 1983), 141.
2. Ibid., 141. By "Intentionality" Searle means that most mental states and speech acts have a reference or are about something. "Direction of fit" can go one of two ways. A mental state or speech act has a world to word direction of fit if it attempts to match the state of the world. Assertives such as "The grass is green" have such a fit. If, however, a mental state has world to word direction of fit, then these states go the other way. Now the world is supposed to be changed to match the mental state or speech act. An example would be "Shut the door." The door in this case is open (the world) and the speech act's purpose is to change the world so the door is shut. By "conditions of satisfaction," Searle means, roughly, what conditions have to be true, correct, appropriate, etc., for a speech act to be issued correctly.
3. Ibid., 143.
4. Ibid., 143.
5. Ibid., 143–144.

SIX
More Empirical and Normative Contexts

ANOTHER VIGNETTE

Thus far only some things have been said about what might be found in our memory banks with respect to problems found on the speech activity level. The focus instead has been mostly on the content material found there in individual speech acts. But our memories are more than information sources of (empirical and normative) individual speech acts. Our memories also have things to say about how we come to gather and organize all these sources so that we can deal with whatever problems we face. They are also about speech activity (discourse).

This process becomes clearer when we imagine a scientist at work. Suppose she and her team of scientists have been at work researching why some people have trouble sleeping and now she is ready to report the results of her endeavors to other researchers. She issues the master speech act to her audience: "I will present to you the results of our sleep research using drug xyz." She reports that she has completed a double-blind study comparing the effects on sleep using xyz and an older drug uvw. She also reports on how some subjects received more of one or the other drug than others, how her team measured different kinds of sleep (e.g., rapid-eye movement versus other forms of sleep), how large her samples are for each group of subjects, and what kinds of instruments her research associates used to conduct their research.

As described, it is evident that our researchers are awash in knowing-that contextual considerations. As experienced researchers, they had to know about other studies that anticipate their own research. They had to know about the chemistry of the drugs they are testing, the personal characteristics and medical condition of their subjects, what equipment was needed to conduct their experiment, and even know where there are facilities adequate to their needs for conducting their research. In addition, they know what each member of the present team knows and does not know, whether they need another medical doctor on their team, and so on.

All this and other knowledge helps our chief researcher make sense of what she reports, but she couldn't possibly have it all stored in the foreground portion of her memory bank. Luckily, she and her team are also awash in knowing-how considerations. They know how to conduct a double-blind study, how to do statistical measurements, how to make careful and objective observations, how to construct proper consent forms, and how to make measurements with the instruments they are using.

In addition, researchers must know how to construct theories, although "must" may be too strong here. After all, some researchers do not engage in theorizing. Rather their projects may only produce relationships between variables, look for laws, and let it go at that. Other researchers are concerned to explain why certain data came out the way it did, and the reasons why the laws were revealed.

Assuming our researchers are interested in theorizing, answering these kinds of why questions, then their know-how might very well yield a theory that explains certain sleep patterns in terms of neural structures. Let us assume, as well, that our researchers have this theoretical know-how and actually develop a theory that seems to explain their results. Well, if they came up with a theory, what have they done and how did they do it? First, they relied on their own memory bank on how to theorize. Second, they developed their special theory on the textual level. In part, they managed to do this by inputting large segments of the data that they uncovered in their research. Third, they may have found it necessary to use their imagination by, for example, employing an analogy. Whether that analogy came from their memory banks or not, it helps them pull together what would become their theory. Having done these, and perhaps other things, the theory would, presumably, be formed enough for the director to give her report.

Let us assume that is what happened. Let us also assume that the report is well received and is later published as a major article in *The New England Journal of Medicine*. Once that happened, the article (or at least parts of it) entered the memory banks of those few who read and understood it. And once that happens, other researchers came to treat the article as a source of knowing for their own research.

But now let us assume that a second group of researchers used a different methodology, arrived at different conclusions, and issued a different report. As their report was also well received, it was also published, and became part of the memory bank for those who read and understood it. However, there is a problem here. Receiving mixed signals from these two reports, other researchers will wonder to which signal they should pay attention. Fortunately, the problem is not serious because, in science, there are standard steps that can be taken to resolve it. Minimally, the new researchers can read over the results from both studies and then possibly discover an error in one that suggests a preference for the other. Beyond that, they can look at both memory entrants, and then may choose to design a new study.

Let us suppose that the new study supports the original study, thereby seriously undermining the credibility of the second. Once published, the new study enters the memory banks of the specialists in the field of sleep research. The original study stays in most of their memories but stays there now with a flag of approval attached to it. Presumably the second study (the flawed one) will stay in the memory bank of some of the researchers but stay there now with a red flag attached to it.

NORMATIVITY AGAIN

Indicating that our memory banks can speak with two voices in the realm of science naturally suggests that the same thing will happen in the normative domains as well. This suggestion might be especially expected in the ethical portion of the overall domain. Why? Because when we enter the process of reason-giving, we sometimes are forced to give reasons to justify the lower level reasons. Eventually the process reaches a level of justification where we are presented with a grand ethical and/or meta-ethical theory. Such a theory purports to explain or justify all (or most) ethical issues, and to do so in a consistent manner.[1] It also purports to do its work in such a clear and organized manner that it is *the* privileged candidate, that is, the one that is considered to be "the best of the lot."

Unfortunately, in ethics most of the theorists (philosophers, theologians, and others) have a tendency to claim that their theory is *the* privileged one. As we have seen, that is not a serious problem if, like science, a recognized set of justification procedures exists to select one of the candidate theories as the better, or only correct one. But there appears to be none in the ethical domain. It is not obvious that there is a procedure, or a set of procedures, that can objectively compare the merits of this theory or that one so as to mark one of them as "the best."

Now let us suppose that, at a very high level of abstraction, the memory banks of professional ethicists, and a few others, hold a collection of four major theories. Or, that they hold variants of each. Let us also suppose that there is no obvious privileged theory in this collection of theories. For those intending to engage in reflective or critical thinking on some major ethical question (e.g., euthanasia, war, dealing with the environment, distributing the goods of society in a fair and just manner) the proliferation of available theories poses a problem about how to proceed.

Should those engaged in such thinking put aside dealing with their ethical problems for a time, and instead try harder to directly achieve theory non-proliferation on the theoretical level? I am assuming for now that this approach would cause frustration. Philosophers have been trying to present us with the "right" theory for some time now. Among specialists in the field, the consensus is that none, so far, has found it. Each theory presented as "right" seems to face continuing objections from critics.

How else might they respond? These moral specialists could deal with their moral problems by appealing to all the basic ethical theories as a single package: one form of virtue theory, one form of Kantian theory, one form of utilitarian theory, one form of contractualism, and, perhaps, some other theory. Would chaos result from this approach? Consider the possibility that the theories might pull in different directions, that they would pull together in the direction of more uniformity, or that they might give forth the same answers, at least on some issues. For sure, whichever direction was taken, the process would be terribly complicated and tedious.

There is one other reason why most of us would not move toward the deal-with-every-theory option. Most (or at least many) of us develop a preference for one or the other major theory early on in our lives. And that preference gets recorded in our memory banks as the preferred theory. How that happens needs to be answered by social and perhaps biological scientists[2] but, without question, it happens. Some of us are inclined to Kantian-like thinking that emphasizes duties and de-emphasizes consequences, some move toward the opposition and favor consequences and place duties in a secondary position. Others are virtue oriented, and probably a few others are oriented to contractualism. So whichever orientation we have, we will, in the end, think about whatever moral problems concern us with a bias for our theory.

It should be clear that moving in a biased way is not necessarily a matter of choice. Our strong and biased academic background could have biased our own thinking. But I speculate (again) that many (some?) of us find ourselves at home with our favorite theory well before we even thought of it as a theory. And being at home here very likely means that, somehow, we came to possess skills in dealing with ethical problems that fit one of the major theories more than the others. In other words, our preferences are more a matter of knowing-how rather than knowing-that.

The "knowing-*how*" preference that we tend to develop has its costs. It tends to leave us blind at times to the proposition that other theories deserve consideration. In its extreme form, knowing how to use one theory keeps us from even seeing others. Not even giving thought to the possibility that their theory has competition, the holders of the theory might very well think dogmatically: "Our 'intuitions' are the right answers to our ethical problems, and that is the end of the matter."

A related response to the "two (or more) faces" problem is possible. Those who have studied ethics academically will be aware of the theories competing with their favorite one. Many of these individuals will be convinced that reason or other intellectual powers that they possess demonstrate that their own theory deserves to be privileged and the competition's theories buried. For them the two-faces problem is only apparent and not real at all.

However, there are other ways to respond to the problem. Already, considering that there is more than one way suggests complexities are found in our memory banks (and the context) not anticipated before. These difficulties and complexities will inevitably lead different thinkers to collect different thoughts in their memory banks. As we have seen, some of the convinced will have one theory privileged so they will see real-life problems only, or mainly, through the lens of their own theory. Others, who are not convinced that they possess the privileged theory, are likely to have a memory bank that looks a lot like many ethics textbooks. There will be a place in their banks for Kantian-like, utilitarian, virtue, contract, and perhaps religious theories. Some of the holders of textbook-like memory banks will not know which theory to employ to help them deal with their ethical problems. Another possibility is that those who follow the textbook format will argue that one theory works best for one kind of problem, while another works best for a different kind. In any case, on the theory level, there will not likely be much general sharing of presumptions. Of course, if Derek Parfit is right, extensive sharing will eventually emerge once some of us work our way through complicated philosophic arguments that show how the major theories merge into one.[3] In the meantime, my memory bank is not likely to look like yours, and yours is not likely to look like mine or theirs. This means that for the foreseeable future, agreement over moral issues will be more difficult to achieve than agreement in the sciences.

NOTES

1. Nick Fotion, *Theory vs. Anti-Theory in Ethics: A Misconceived Conflict* (New York, Oxford, and London: Oxford University Press, 2014), Chapter 3.

2. S. Matthew Liao, ed. *Moral Brains: The Neuroscience of Morality* (New York: Oxford University Press, 2016). Articles in this volume are by the following authors: S. Matthew Liao, Jesse Prinz, Jeanette Kennett and Philip Gerrans, James Woodward, Joshua Greene, Julia Driver, Stephen Darwall, R. J. R. Blair, Hwang Stuart, F. White, Harma Meffert, Ricardo de Oliveira-Souza, Roland Zahn, Jorge Moll, Molly J. Crockett, Jana Schaich Borg, Guy Kahane and Walter Sinnott-Armtrong.

3. Derek Parfit, *On What Matters*, Volume One (Oxford: Oxford University Press, 2011), 411–419.

SEVEN
"Logic" and Contexts

"LOGIC"

In his "Discourse Ethics: Notes on a Program of Philosophical Justification" Jürgen Habermas uses expressions such as "presupposition" and "transcendental" to talk about what I am discussing under the heading of "context."[1] Presuppositions seem to be beliefs, attitudes, and tendencies of behavior that we are committed to when we use language (i.e., issue texts). Some of his interests in presuppositions are pretty basic in that we don't just choose them haphazardly or even choose them after we have thought about them for a while. Rather, they are in some sense to be specified below, impossible to avoid if we want to use language at all. This makes Habermas's presuppositions discussed in his paper different from most of those discussed in earlier chapters of this work. In these chapters, the presuppositions might have been put there by the speaker or might have acted as commitments if the speaker had chosen to work in a specific linguistic domain (e.g., in science).

In contrast, some of Habermas's presuppositions apparently are taken by him as avoidable only if the speaker fails to speak. We will see shortly if that is the case. About these necessary presuppositions he says:

> It makes sense to distinguish three levels of presuppositions of argumentation along the lines suggested by Aristotle: those at the logical level of products, those at the dialectical level of procedures, and those at the rhetorical level of processes.[2]

The logical level of products he also calls the logical-semantic level. I call it the level of the speaker's logical duties. The sample list of these duties he offers us is as follows.

(1.1) No speaker may contradict himself.

(1.2) Every speaker who applies predicate F to object A must be prepared to apply F to all other objects resembling A in all relevant aspects.
(1.3) Different speakers may not use the same expression with different meanings.[3]

Several comments about these principles are in order. First, they undoubtedly apply beyond the field of Discourse Ethics. Obviously, for example, rule (1.1) applies to all (or almost all) "language games" not just to ethics. No doubt, Habermas would agree. Even rule (1.2), which comes off as the universalizability principle in ethics,[4] has a wider application than the language game of ethics. The same can be said of rule (1.3). It also applies to almost all "language games" or domains that we can think of. Second, and similarly, he says that the rules he presents us are in place when we are engaged in argumentation. But these rules also have wider application as when we simply are talking, reminiscing or what have you. Normally, we don't want to contradict ourselves, for example, even when we are telling fictional stories. Indeed, for the purposes of this project concerned with canvassing a large part of the context (and its memory base/bank), it would be inappropriate to limit the analysis to argumentative discourse in ethics.

Third, Habermas says that his list of principles or rules contains only examples. Well, what additional examples fit in this category of speakers' logical duties? Certainly, the logical form of argument that we all call *modus ponens* does. If the claim "A" is true and if the conditional claim "if A then B" is true, it follows logically that "B" is true. But if *modus ponens* belongs on our list, then so does *modus tolens, disjunctive syllogism, hypothetical syllogism* and a whole host of logical principles that many of us have studied in basic logic courses.

Fourth, there is a need to comment on the claim that following these principles is unavoidable. In this connection, note how Habermas characterizes "communicative."

> I call interactions *communicative* when the participants coordinate their plans of action consensually, with the agreement reached at any point being evaluated in terms of the intersubjective recognition of validity claims.[5]

In effect, Habermas is saying that someone like Garbo, the famous double agent of World War II who deceived the Germans into believing that the 1944 invasion of Europe by the Allies would take place at the Pas de Calais area of France, was not communicating with the Germans.[6] He may have sent thousands of messages to the Germans, and he may have received thousands in return. Further, the Germans may have communicated with him since they were presumably following the rules of logic and all of the other communication rules (see below) that Habermas tells us about. But Garbo was not communicating.

I submit, as part of the third comment on the speaker's logical commitments, that Habermas is engaged in a bit of a stipulation concerning the meaning of "communicative," "communication," etc. Indeed, Garbo is communicating with the Germans even though he may be violating some conversational rules. This means that the necessity associated with following these rules applies only if a speaker is following normal usage (I use "normal" for want of a better word). Now Habermas may have his reasons for stipulating as he does. But for our purposes, giving "communicative" a narrower meaning is harmful. As was just noted, the goal here is to give an account of as much of the context and its associated memories as possible. In contrast, Habermas's stipulation cuts off the analysis by suggesting that the context is structured in such a way as to disallow that a conversation is taking place if one or more of the conversational rules has been violated.

Taking then the broad meaning of "communicative" ("communication"), here is how Garbo's communicative process with the Germans worked. Garbo sent many texts to the Germans. These texts were received, read and then filed away into their memories and their cabinets. Gradually their memories (and cabinets) became loaded with misleading information. When, a few days after the Normandy invasion started, Hitler and Field Marshal Gerd von Rundstedt (the German military leader on the Western front) decided to move the *panzers* south to the Normandy area. At this point Garbo sent a message (messages?) saying that Normandy is a diversion and the real invasion will, indeed, take place at the Pas de Calais. Thus, he recommended that the Germans turn back the *panzers*. And the Germans turned back many because what Garbo told them fit into their (false) memories that Garbo himself put into their brains.

So Garbo was communicating with the Germans, and effectively. After all he seems to at least be following the speaker's logical duties. The Germans wouldn't have paid any attention to what Garbo was telling them if his messages made no logical sense. But, for Habermas, Garbo wasn't communicating because he was violating other transcendental rules.

Habermas calls his second "catalog of rules" procedural. It receives the formal title of the dialectical level of procedures. I call these rules the speaker's communication duties.

(2.1) Every speaker may assert only what he really believes.
(2.2) A person who disputes a proposition or norm not under discussion must provide a reason for doing so.[7]

Comments are in order for this set of rules as it was with the first set. First, it is obvious that while Garbo is texting regularly, possibly furiously, he is violating rule (2.1). This rule is an example of what those who write in the speech act tradition call the sincerity condition.[8] In "normal" communicative settings it is assumed that the speaker believes what he utters when issuing an assertive speech act (e.g., "Churchill was at the party"). However, if the speaker utters a commissive speech act (e.g., a promise) the sincerity condition changes to intent. It is assumed when the commissive is issued that the speaker (truly) intends to do (or at least will try to do) what she promised. Similarly, there is a change when a directive is issued. Now the sincerity condition is a wish (or want or preference). So if General Montgomery says, "Shut the door," we assume that he sincerely wants it shut. Accepting a modified version of (2.1) to include different sincerity conditions for different basic kinds of basic speech acts, it is even clearer that Garbo's way of communicating with the Germans is "big time." Specifically, he is violating the truth telling rule extensively and systematically. But he is indeed communicating, if we mean by "communicating" something broader than what Habermas means by that term.

As to (2.2), Garbo, could follow that rule without worry. What it tells us is that if he, or anyone, wishes to change the subject under discussion, he needs to justify the change. Use of "discussion" makes it clear that Habermas is thinking exclusively now of how discourse works. In contrast, (2.1) is at times a speech act rule.

One can add to Habermas's list of the speaker's communicative duties rather easily. On the speech act level, there are the following rules:

(2.3) Every person should employ a language that he shares with the hearer(s).
(2.4) Every person should speak clearly (so he can be heard).
(2.5) Every person should write clearly (so he can be read).
(2.6) Every person should speak clearly (so he can be understood).

These rules have their correlates with discourse whether it is argumentative or not. In addition, there are the following discourse rules or principles to consider:

(2.7) Every person should organize his discourse, perhaps by issuing master speech acts (so as to aid his hearers in following whatever thoughts he is communicating).
(2.7a) Every person should follow the lead set by the master speech acts (whose purpose is to control the direction the speech activity hopefully will take).

In understanding the status in the context of all these principles or rules, it is important to contrast them to (1.1). (1.1) tells us "No speaker may contradict himself." In fact, a speaker can contradict himself and might do so if he thinks he can get away with it. But, more often than not, he pays a price when he does so. Others hear what he says so it looks like he is "communicating." But what looks to be so isn't so since most of his hearers can't make sense of what he is saying (or writing). In contradicting himself, he is saying something and then taking it away. So (1.1) is a rule one must follow if he wants to reach most of his hearers. Some of the (2.x) rules have this same status. If, as with (2.3), a speaker uses a language unknown to the hearer, communication will be aborted from the very beginning. The same is true with (2.4). If the speaker whispers to a large audience (without the aid of a microphone) communication will fail from the start. It will also fail, of course, if the hearer is deaf and/or as a reader, is blind.

Some of the (2.x) rules, then, have the same status as (1.1). But others do not. (2.6) tells us to speak clearly if we wish to be understood. If the speaker is extremely unclear then, once again, we have a rule like (1.1). However, there are degrees of lack of clarity so that understanding can come about but does so with great difficulty. So now "must" means something like if you want to facilitate communication, speak clearly. "Must" now means "conditional necessity." With (1.1) "must" is not conditional in this sense.

Now let us return to the hearers. They too have rules that they must follow. Here is a suggested list of the hearer's communicative duties, as well as a list of his logical duties:

(2.11) A person must be able to hear (or read). This is not really a duty but a condition that the hearer cannot correct or, perhaps, correct with great difficulty.
(2.12) A person must pay attention to what the speaker says (or writes).
(1.11) A person will criticize or not allow the speaker to contradict himself.
(1.22) A person will criticize or not allow the speaker to violate rules (1.2).
(1.33) A person will criticize or not allow the speaker to violate (1.3).
(1.44) A person will criticize or not allow the speaker to violate any of the rules of logic covered under (1.4).

One final comment about the (1.x) and (2.x) rules is in order. These rules mostly fall under the heading of knowing-how. We can articulate and thus grant them knowing-that status, as is being done in this chapter, or as might be done in a logic class. However, most of the time these "rules" represent habits of behavior. We exhibit our knowledge of *modus ponens*, for example, simply by arguing in accordance with that rule of logic, but without necessarily consciously following or reciting it.

Chapter 7

MORE "LOGIC" — SORT OF

In his *Discourse Ethics* article, Habermas offers us one other set of rules that are presupposed when we communicate with others, as he says, in an argumentative setting. He calls this set the rhetorical level of processes. I call them the speaker's community rights and the hearer's community duties.

(3.1) Every subject with the competence to speak and act is allowed to take part in a discourse.
(3.2a) Everyone is allowed to question any assertion whatever.
(3.2b) Everyone is allowed to introduce any assertion whatever in the discourse.
(3.3) No speaker may be prevented, by internal or external coercion, from exercising his rights as laid down in (3.1) and (3.2).[9]

Viewed as rules governing formal or semi-formal argumentation (e.g., as in a debate or at a professional conference), these rules are perfectly acceptable. They have a comfortable home in at least some of our memory banks. However, in some settings, these rules do not seem to fit at all.

Consider an argument between a well-informed physics professor and a group of Introduction to Physics students. Hardly anyone would argue that her students should be denied their say about certain laws of physics. Students will (sometimes) learn more if they have the opportunity to speak up. But achieving the goals of education is one thing and reaching the goal of a rationally motivated agreement is quite another. In time, "rationality" may require that the professor quiet her students, perhaps even coerce them, in order that they all reach a rationally motivated agreement. That is, after a while, unlimited student talk on the subject at hand may hinder the process of thinking clearly and learning. Thus, Habermas's (3.x) rules do not necessarily apply in some settings. One should also understand that if one extended the application of these rules beyond argumentative speech, these rules may not apply at all. There may be a time for communication between a parent and an unruly teenager to be one sided, even coercive. Similarly, communication that may or not include argumentation may be one sided if one participant is a general in the army and the other is a newly-minted lieutenant.

Pointing to a distinction highlights another problem with Habermas's (3.x) rules. His process rules tell us how we should proceed in deciding what to do. His list of principles is a wish list of what should be in our memory banks. But these rules say nothing about what actually is in our banks. Some (a few) may actually have a place in our banks. But many, likely, do not. So it makes sense to ask whether some decisions should be made in such a way so as to mimic his conversational process? Is everyone, by voting, going to help make a proper Habermas-like decision? Such a participatory democratic decision may be one way to go (perhaps in some political settings). But at the other extreme, we may allow every competent speaker to participate in the discussion but have only one person making the final decision to act (as in a military setting).

As characterized by Habermas in his "Discourse Ethics," it is not clear whether he wants community-based democrats, dictators, or some intermediate group making the actual decisions. For our purposes what he wants is not terribly important. Rather, it is important that his procedures are compatible with more than one decision process. We can assume, then, that each option (choice) is registered someplace in our context memory banks. In other words, if someone were to inspect all the crevices of our minds she would not find a single answer (e.g., presupposition) to the how-shall-we-decide question, but many.

Some thinkers might not appreciate the opportunity available to them to examine all the options to see which one might be preferred. This lack of appreciation may come about, indeed it is likely to come about, since these thinkers may not have examined all the corners of their memories. Instead, they might see, for example, that the communitarian option is necessarily presupposed in their reflective thinking on some major problem or other. For them, then, an examination of their own presuppositions would be part of the process of justifying their answer to their problem. "See," they say, "we have delved deeply into our presuppositions (which seem necessary to them) and we have, indeed, given a complete account of what we believe and why."

But if this line of thought is correct, ideas of a complete or single-decision account is probably flawed. These single-account thinkers usually hold onto a further presupposition that an examination of their memory banks yields only one answer in the end to whatever problem concerns them. At times, of course, our memories contain a single answer (to any one problem), and, by consensus, we sometimes identify that answer as correct. But it doesn't necessarily work that way in many settings. The single-answer account underestimates the complexity found in our memory banks. One doesn't want to say that our memories contain everything since we know that more can be added to them as, for example, new technologies come online. But the banks of many (most?) of us are rich enough to allow for many answers to one's questions.

In closing the discussion of Habermas's principles of discourse ethics, one should not forget that his discussion is not concerned with the issue of how we as language users actually think. He is not telling us about what is in our memory banks. Instead, Habermas is giving us a portrait of ideal thinkers and ideal users of language. Some users may actually have his principles in their memory banks and have them fully employed. But some may not. More likely many will have a mixed bag of some of his principles in their memory bank. Habermas, then, in writing his article, is playing the role of a logical (conversational) prescriptivist. He is attempting, as all logic teachers do, to change most of our memory banks so that in the future we will think more logically. To that attempt, we are tempted to say, "good luck."

NOTES

1. Jürgen Habermas, "Discourse Ethics: Notes on a Program of Philosophical Justification," in *Moral Discourse and Practice: Some Philosophical Approaches,* ed. by Stephen Darwall, Allan Gibbard, and Peter Railton (New York, Oxford: Oxford University Press, 1997), 287–302.
2. Ibid., 294.
3. Ibid., 294.
4. R. M. Hare, *Moral Thinking: Its Levels, Method and Point* (Oxford: Clarendon Press, 1981), 87–106.
5. Habermas, *op. cit.*, 287.
6. Garbo's real name is Joan Pujol Garcia. There are many sources of information about him on the Internet.
7. Habermas, *op. cit.*, 295.
8. John Austin, *How to Do Things with Words,* Second Edition, ed. by J. O. Urmson and Marian Sbisa (Cambridge, MA: Harvard University Press, 1975), 15. Originally published in 1962.
9. Habermas, *op. cit.*, 295.

EIGHT
Still More "Logic"

"LOGIC"-LIKE FEATURES

When we urge someone to: (1) "Identify all the possibilities," it sounds as if we are encouraging this person to think logically, or at least to think in a logic-like fashion concerning an empirical matter on which he is working. Along this line, we can say a bunch of other things pertaining to doing empirical work.

1. When you make observations (in an experiment), repeat them when you can.
2. Make your observations in clear light, and in a quiet setting.
3. Pay attention to the observations of others.
4. Pay special attention to technically trained observers.
5. When you are generalizing from a sample population make sure that your sample size is large enough and unbiased.

There is also a list for working in the normative domain. Here is a sample that, to some extent, overlaps the empirical domain.

1. Do the best you can to give (or listen to the) reasons to support all sides of an issue.
2. Gather the relevant facts carefully before making final judgments of right or wrong.
3. Be suspicious of your judgments if you find yourself in a "conflict of interest" situation.
4. Be suspicious of other peoples' judgments if they are in a "conflict of interest" situation.
5. Make your decisions when you are in a calm and quiet setting.
6. Do not make your decisions if you are fatigued, emotionally disturbed, on heavy medications, drunk, and the like.

Both of these sets of "guidelines" can be, and are sometimes carried and shared in our memories either as habits or as rules. Those who constantly make important decisions probably do so in "automatic" mode. Thus, all of these rules or guidelines for them are not cited. They are instead followed without thought. Others less used to making such judgments will feel more comfortable citing the rules.

MISCELLANEOUS

At first, so called everyday sayings do not appear to form part of the memory bank. Indeed, they are cited so often (or often enough) to appear to be more associated with the text. That is why they are called sayings. Consider the saying "The best is the enemy of the good." When cited, the listeners will hear that as part of the text. If they had never heard the best/good saying before, they might be puzzled as to why the speaker cited it at just that moment in the discussion. Their puzzlement will be there even if they understand that "The best is the enemy of the good" means something like "You are better off trying to achieve a high but reachable standard of performance than the unreachable highest standard." Even with this understanding in hand, they need to be told that the context surrounding what was said has to do with, for example, designing cars; and that a company makes a mistake when it spends much too much money "In pursuit of perfection," as Lexus is prone to tell us in their television ads.

In the first hearing of this best/good saying, then, it is clearly not part of the memory bank. In time, days after the first hearing, the saying may have been completely forgotten. But if remembered, it will have migrated to the bank and thus be in position to play a silent role in a vote of a board member of an automobile manufacturer. The board member might now vote "no" to a proposal to spend still more money on an exotic new car, whereas before hearing the saying, he might have voted "yes."

Consider some other sayings. "You are barking up the wrong tree," "Pride goes before the fall," "All good things come to an end," "Grass is (always) greener on the other side of the fence," "A fool and his money are soon parted," "You've put the cart before the horse," "Just do it," "He who hesitates is lost," and "Put your shoulder to the wheel." Each is present in our shared memories of many people and so is ready to be used in the context when the occasion arises. Each saying can be taken out of our memory bank, quickly given the status as part of the context, and then raised to the level of the text if necessary. And after it is used, it most likely will return to its place in our memories waiting to be used again at some later time. Like the best/good saying, the effect these sayings have on the text may not be caused by a deliberate decision of the speaker to invoke them. If the setting is right, each saying may affect what is said in the text silently.

LOOKING THINGS OVER

These sayings, the logic-like recommendations cited earlier in this chapter, and the more formal "transcendental" principles (e.g., cited by Habermas) perform radically different functions from those performed by the memory banks cited in Chapters 1–7. The memories from those earlier chapters provide information of all sorts either in the form of knowing-that or knowing-how, and tell us about things either descriptive or normative in nature.

We assume that this knowledge, at least the vast majority of it, is correct. If we had found in the past that very large segments of knowledge are unreliable, wrong, and/or false, we likely would have given up on using language. Although reliable overall, we know from experience that some texts can go wrong and some portions of our memory can go wrong as well. When they do, for whatever reason, tools are needed to identify what has gone wrong and then, if possible to fix the wrongs. Concerning memory bank, the tools for fixing things look a lot like the logical and communicative principles Habermas offers us, and other logic-like principles.

All these logical and logic-like principles have two distinct roles to play. Thus far the discussion has focused on remediation, on fixing things when they go wrong. But these principles also have an expansive role as well. When speakers use logical principles (along with assertives, directives, etc.) to help yield conclusions not articulated before, they are taking steps toward expanding their memory bins on the speech activity level. When they use these principles (along with assertives, directives, etc.—all in their memory banks) to assess other memory bank items already in the bank, they are engaged in remediation work.

Many people (perhaps most of them) do very little expansive work. On the face of it, it appears otherwise. After all, these many people are constantly adding memory items to their banks about themselves, their friends, their enemies, and media stars. But no matter how many of such items are added, the texts they generate cannot help but be narrow as to content. Little or nothing is found in their banks about national events, world events, important ethical issues discussed and presented in any of the media outlets, important scientific engineering developments in space or medicine, and certainly nothing much about matters relating to logic.

We can appreciate the nature of this content narrowness by observing some people who are truly memory-bank expanders. To be sure, they talk about themselves, but they also read books, listen to national and international news programs, read newspapers such as the *New York Times* and the *Washington Post,* and access all kinds of media to help give them a "world" perspective. Some in this group are also expanders because they are specialists in medicine, chemistry, or some other science. Other specialists are expanders in economics, political science, and so on.

Given these vast differences in "life style," vast differences in how much expansion in their context memory are to be expected. People may have been created equal, but later in life their memory banks, for better or for worse, are clearly not equal.

Equality is missing as well when it comes time to do remedial work. If we are not well-rounded with memory-bank extenders, there is little likelihood that we will do effective remedial work. To do such work we need enriched memory banks on the logical and logical related levels. But before that, we need to have the incentive to do remedial work. Those of us who are content to expand our memory bank only on local issues about ourselves and friends are not likely to even think about remediation. Of course, we can be reminded. Recently (2016) Cadillac had a television ad asking its viewers to "Challenge your preconceptions." Presumably those planning to buy or lease an upscale automobile are thinking positively about Mercedes-Benz, BMW or some other high-end foreign machine, but not about Cadillac. But using "challenge" suggests that remediation is not easy to do. Again, then, local thinkers (I am speculating here) are not likely to be looking for challenges.

Even if one likes challenges, local thinkers are not going to be able to generate challenges of any kind since they do not have the logical tools to get the job done. Their memory banks are relatively empty. But even if the bin is full of logical and logical-like principles, they are not necessarily going to use these principles to arrive at rational decisions.

To explain why, it is helpful to digress for a moment in order to say a few things about the structure of our memory banks. Up to now, we have seen that our banks carry an extremely wide variety of memories. Some of these are factual (assertive) while others are more or less infected with normativity. Now in this and the previous chapter, I posited that there are logical and logical-like principles in the bank as well. In addition, we learned that some are social in nature while others are more personal. Still more, we learned that within any one domain, as in ethics, some rules and principles are shared by all, while others are not. Still more, we learned that the sources of these principles vary. Some principles, like the law of non-contradiction and the ethical principles that form what Beauchamp calls the common morality, may be built in our nature. Other principles are learned in settings shared by all while others are learned, at the opposite extreme, on the basis of individual experience. Given this variety of natures and sources of our memories, it is highly likely that the bank will contain some entries that are consistent with one another, while others are not. In short it would be a great surprise if the jumble of principles, rules, individual claims in our (individual) memory banks are not frequently in conflict with one another.

We are in position to answer the why-not question. Why might John (or anybody), even though he has the habit of acting in accordance with reason, not follow reason? Quite simply, because he has other habits that conflict with his habit to reason. He exercises his reason when making non-life saving decisions. So he challenges his presuppositions that weigh against buying a Cadillac by using logic. But when he deals with weighty decisions he resorts to prayer. Others like John act similarly. They by-pass reason, and instead study tea leaves. Still others appeal to some authority by asking themselves "What would Jesus (Ghandi or X) do?"

Put differently our individual memory bins are not the creations of a brilliant philosopher's rational imagination. Rather, many persons, with varied agendas, put it all together so that large portions of the bank are a mess for most of us.

But not a total mess. After all, the logical principles discussed in the previous chapter are in position to help. These principles are not chosen by individuals. Coercively they bring about sharing. To be sure, some of us are mentally challenged and so don't recognize the impact of these principles. But acceptance of these principles is so widespread (among peoples both now and in the past) and is so every-day pervasive in both language use and in our thinking, that order in our memory banks begins to seem possible. In addition to being widespread and pervasively present in our banks simply as principles, these principles work for us by way of helping bring order to a variety of life's problems (e.g., in ethics and science), and thus bring even more order to our texts.

As we saw in Chapter 4, further aid toward avoiding messiness in our memory banks and in our texts can be found in the ethics domain. Recall again the role played by the common morality. That morality does not seem to be forced on us the way logic is, but it still seems hard to avoid. What is the alternative to the "Don't harm" principle? What is the alternative to harming, perhaps, killing one another? And what about telling the truth? Should we adopt a principle that encourages systematic lying? Very few would choose principles that go against those found in the common morality. Many different philosophic and religious stories can be told to justify the common morality principles. However, justification is not the task being undertaken here. Rather, it is giving an account of a segment of the memory bank that helps to avoid chaos. It was said that there is much chaos in each of the banks that each of us possesses. Not surprisingly, order is somewhat restored by the logic domain and also by the sub-domain of the common morality.

There is another order producing sub-domain. In Chapter 5 it was called basic beliefs. That sub-domain contains what we come to believe from everyday experience. You believe, for example, that the ground you walk on between your house and the neighbor's is solid. But your belief is not expressed cognitively by the thought (and language use) "The ground between our houses is solid" but by simply walking there. You are only thinking of your duty to return the vase you borrowed a week ago to your neighbor. Similarly, one does not remind oneself when talking to a group of citizens that they are all real creatures and not sets of abstract philosophical constructions. Rather, one presupposes that they are real creatures, all the while talking to them about why Hillary Clinton lost the US presidential election of 2016. There are lots of basic beliefs around, none of which are normally coded into the memory bank cognitively. Some sound more "meta-physical," others more physical. Under the former heading are mountains, trees, lions, etc. All are taken to be real. Under the latter are beliefs that rain is composed of water, that fire is hot, that weeds are hard to get rid of, and that one cannot help but make noise as one moves over ground heavily laden with autumn leaves.

There is a wide variety of beliefs in our memory banks that are prone to create conflict and chaos, but there is also a wide variety of stabilizing beliefs present. It appears that neither side is strong enough to dominate our language use and/our thinking. The chaos or messiness is not strong enough to sustain skeptical arguments so that we can say or show that we know nothing. Yet the power of logic, scientific method, the common morality, etc., is also such that it cannot settle all our disputes. In short, there seems to be a kind of balance between the forces of order and disorder.

But what kind of claim is that? It is tempting to think of it as an empirical claim. For a time, some young people have tendencies toward living disorderly lives and then, as they mature, move in the direction of more order. Thus, they achieve a kind of balance in their memory banks over time. So, as they change, their memories move from disorder to order as well. However, such an account does not make much sense since there are others who live in disorder all of their lives, while still others live in perpetual order. There is, as we have seen already, apparently too much variation in what is contained in the memory bank to allow for balance to play an important empirical role in these matters.

Once again, then, what kind of claim is it to say there is balance in this discussion? Maybe it is some sort of normative claim. Balance, that is, order mixed with disorder, some might think, is what we should try to achieve. Order alone might be thought of as not exciting, and disorder alone too chaotic. The balanced life even sounds a bit romantic.

Perhaps a case can be made for normativity here. But that doesn't seem to be the direction the discussion has been taking. It sounds more descriptive than normative in nature. But as a descriptive account what can it be if it is not empirically descriptive? The only option left is that it is a descriptive conceptual account. Balancing thus must mean the memory bank contains, on the one side, a host of important concepts (e.g., as found in religion) that don't lend themselves to resolution of conflict and yet contains a host of other important concepts that do tend toward resolution. And, further, the reason there is balance is that neither of the two forces is dominant enough to crush the other.

NINE
Expressives and Declarations

EXPRESSIVES

Expressives are a strange lot, so I am treating them separately from the other major types of speech act. Declarations are a strange lot, too, and they deserve separate treatment as well. I'll start with expressives and deal with declarations later.

Expressives are strange in large part because they are extremely varied, and so are difficult to define or even seriously characterize. The best that can be done is class them into two groups, namely, social lubricants and social irritants. An example of the former is "Good morning" issued to a pedestrian walking the other way. If you get no reply, and yet are sure you were heard, we have a true case of a speech act showing up by itself. If the pedestrian passing by responds with "Good morning" we have a case of minimal speech activity.

Working with the latter example first, it is clear that "Good morning" appeals to a rich segment of one's context memory pretty much like more extensive uses do. The realism principle is still at work. We don't greet what we suppose are ghosts. We also don't greet robots, although we might stare at them if we haven't seen one walking by lately. Also, we don't greet the passerby in English if we know that he knows no English. We need a shared language even when the speech activity is ever so brief. We also assume when we greet someone with "Good morning" that both you and your fellow pedestrian will still be human in the morning and into the evening. We do indeed make lots of the same assumptions as we do with other more complex forms of speech act and speech activity. However, there is at least one difference. We have no need in this brief contact to appeal to master speech acts. So, although the context in our greeting example is complicated enough, it is, to a small degree, simpler than other uses.

Actually, there is another difference that needs noting. Because the greetings contact is so brief, it is possible to bring it about without using one's natural language. If the two passersby are yards apart and if there are loud noises from nearby yard machines, the two pedestrians may simply wave good morning (or hello) with their arms. Of course, one can supplement the wave with a smile.

Here is a somewhat more complicated but common example of communicating without using one's natural language. A driver is trying to move from a parking area onto a very busy street. It all seems hopeless until a second car stops and thus creates a space for the first driver. Given the driving culture of the community, the first driver sees that the second driver is inviting her to move into the space he has created. In effect, the second is issuing a directive which has within it all the assumptions, presuppositions found in natural-language directives. Taking his cue from the actions of the second driver, the first driver moves into the empty space, rolls down her window and signals the second driver (now behind her) with a wave of her arm. Her wave is taken as meaning "Thank you." Notice, as an aside, how the same (kind of) wave has the meaning of "Good bye" at a dockside as the ship is leaving. What makes signaling appropriate in each case is the context. Driving on the road has its own practices as does watching a ship departing. Even though in both cases the situation is such that ordinary conversation is impossible, we find other ways of communicating with our fellow humans (e.g., by tooting one's horn). And even in these sorts of situations, we find that a (shared) context of one sort or another is fully at play.

Both examples of expressives I have given are social lubricants. In a small way, to be sure, they make life go better. Other commonly used social lubricants are "You have my condolences," "Congratulations," "My apologies." Of course, each carries its own special context. In one an unfortunate event, in another, something has gone well for the listener, and in the third, the speaker has caused an unfortunate event. Other social lubricants include "Good luck," "Good bye," "Have a nice day (or weekend)," "Have a good (safe) trip," "Happy New Year," and for Christians "Merry Christmas." The list could go on.

SOCIAL IRRITANTS

With some irritants, they may not be speech (communicative) acts at all. "Damn it," said immediately after you have dropped your drink on your expensive rug may be an emotive expression most akin to a scream. However, at times, perhaps well after the drink has fallen, "Damn it" may be a genuine speech act. Roughly speaking, the same can be said of the expression "Damn you!" except now one is not speaking to himself.

Most of us are more than familiar with a host of other social irritants. Some of the most famous are "God damn you," "Go to hell," and the milder "Go jump in the lake." The latter expression is especially interesting since, to some, it sounds more like a directive than an expressive. But one needn't necessarily take sides in this discussion. It could be that this expression is both a directive and an expressive depending on the context.

But analyzing first this speech act and then that one is not the primary concern of this study. What is, is the nature of the context that supports these and some other utterances that are clearly filthier. In that connection, it is obvious that social lubricants and social irritants are very much alike. They both appeal to what I am calling the reality criterion (i.e., we take ourselves as apologizing to or swearing at real people), we assume that these people will be around for a period of time, they will not change into apples, the environment in which they live will be much as the scientists tell us it has been for a long time, both the speaker and hearer know how to use the language used by the speaker, etc. There is, in short, nothing very unusual happening among expressives that marks them as different from the other speech acts (assertive, directives, evaluatives, and commissives) examined so far. Perhaps the main difference is that since expressives are largely stand-alone speech acts, some of the principles of logic that pervade speech activity do not apply to them.

DECLARATIONS

Declarations are a "different breed of cat." John Searle has studied and championed these speech acts more than any other philosopher/linguist. In his "Taxonomy" article, he introduces his readers to declarations as follows:

> *Declarations.* It is the defining characteristic of this class that the successful performance of one of its members brings about the correspondence between the propositional content and reality, successful performance guarantees that the propositional content corresponds to the world: if I successfully perform the act of appointing you chairman, then you are a chairman; if I successfully perform the act of nominating you as candidate, then you are a candidate; if I successfully perform the act of declaring a state of war, then war is on; if I successfully perform the act of marrying you, then you are married.[1]

His example of appointing a chairman helps to show how it all works. Smith can be appointed chairman of a committee by Roberts because the latter has been recently appointed to be a regional vice president within The Corporation. It is his new status that gives Roberts the power to hand over other powers to Smith. Presumably these powers are less than those that Roberts possesses. Still the "lesser" authority that Smith has received now allows him to chair a committee to investigate why there is conflict among the members of the shipping department and how this conflict leads to inefficiency. In making the appointment, Roberts will say to Smith something like "I am appointing you chairman of this committee." Searle says that the semantic structure of this declaration is something like "I declare: you are to be the chair." In issuing the declaration, Smith, magically, becomes the chair. Given the social structure of The Corporation, Smith now is the chair. It's a done deal.

But now the question arises, where did Roberts get his authority to make this kind of appointment? The answer is, from the senior regional vice president Quincy. By issuing the declaration "You are appointed (promoted) to the position of vice president" Roberts magically became a vice president. That is a done deal as well. But then where did Quincy get his authority? Obviously, from some higher-up official at The Corporation's central headquarters in Chicago. The most senior vice president in charge of personnel in Chicago put Quincy in his place by using the authority given to him from a more senior official of The Corporation. On it goes, until we get to the top.

Things at the very top are different since there is no one in those higher regions to declare Mr. Adams the company CEO. Obviously one possible way is that Adams became the CEO because he started the company that later became The Corporation. When it came time to hire the first employee then, of course, only he could make that hire. In short, he could have become the CEO without any formal issuance of a declaration.

It is easy to understand how declarations can be thought of as powerful context generating machines. Institutions of every kind (business, government, the military, religious organizations, academia, and so on) just don't work within an already-in-place context. Declarations help them create important parts to that context. Declarations help create the context (memory) in another way besides placing certain people in certain positions. To be sure, Roberts cannot only appoint Smith chairman, he can, if he likes, appoint a whole host of others to be members of the committee. But beyond that, he, or more likely someone higher-up, can actually create a committee. This higher-up person can do this by publicly saying "I am creating a new committee to do the job of preventing The Corporation from being electronically attacked by its enemies." In this kind of setting, declarations are not just supporting existing institutions. They are helping by way of creating them. Realizing that this role can be applied to all sorts of settings, one can begin to appreciate how powerful declarations are in giving shape to our social practices.

Declarations also play a role in the process of naming, a process of great interest to philosophers. Not all naming takes hold because of declaration issuance. One can find him/herself attached to a name without knowing how the attachment came to be. A good example can be found in the naming expression, "Roosevelt's Rough Riders." At first, "rough riders" was used descriptively.[2] Teddy Roosevelt's regiment was composed of a mixed bag of Ivy League football players, Texas Rangers, ranchers, cowboys, Indians, and variety of roughneck types. "Rough riders" was indeed an apt description. Only gradually did rough riders come to be "Roosevelt's Rough Riders" or simply "The Rough Riders." As far as is known, no one blessed Roosevelt's troops by giving them the name history has bestowed on them. In contrast, if one is associated with some Christian or other religious, group it is likely that the name you carry is the one given to you as part of a special religious ceremony. You then carry the name as you grow into adulthood and even after you have "passed."

LESSONS

Here is a list, with comments, of some of what has been learned in this and earlier chapters as to how context can be altered.

1. Even the simplest expressives carry with them layer upon layer of context material brought in from what was originally textual material. These expressives alter the context only in a minimal way. They do so simply by adding them to the memory bank. Having done so, I am more likely to respond with a "Good morning" to someone who greeted me with her own "Good morning" yesterday.
2. Assertives tend to alter the context bit-by-bit. Some assertions are added to the local context, as in Chapter 1, where the boys are talking by themselves about Derek Jeter's baseball accomplishment. But others alter the context more generally such as when they are issued to a large number of people at a conference, in the NY Times, or on MSNBC. The billions of assertions issued every day make it clear that assertions, unlike expressives, are major context changers. Of course, many assertions change the context only temporarily. We say many, or perhaps most, of them are forgotten. Others don't even get the chance of entering and then leaving. They are forgotten even before they had a chance to enter the memory bank. Still many stay there and so they significantly alter the bank by adding items to it.
3. Commissives add to the context in a unique way. They create duties (not facts) which, at first, are thought of as part of the text, but they too eventually settle into the context's memory bank. Their numbers are surely fewer than assertions but the effect they have

on the context is still great because we take many commissives seriously.
4. Directives have a major effect on the context. They are issued in large numbers, but probably less frequently than assertions. Whether directives enter and stay in the bank depends on, among other reasons, who issues them. Directives issued by authorities (presidents, generals, bishops, professors, etc.) get the attention from listeners. These directives thus find their way into the memory bank of many. Most directives issued by lesser figures are forgotten.
5. Evaluatives are issued in large numbers. We rank people, cars, houses, political proposals, almost anything.
6. Declarations are not issued as frequently as are some of the other basic speech acts. But their influence on the context is unique and great. Other speech acts, such as assertives and directives, change the context by adding to and subtracting from information. Otherwise they leave the context alone. They don't alter its structure. They add and subtract what we are calling content to everyone's context. Declarations can do this too. But what makes them important is that they add, and occasionally subtract, to the very structure of the context. They do this only in the social realm (see especially Chapter 3). These structural declarations can create whole institutions (as the U.S. Declaration of Independence did and does) or create parts of institutions (as when an official of a corporation creates a new department inside that corporation). Declarations can alter the context as well by creating practices (such as oaths of allegiance).
7. Master Speech Acts change the context in a peculiar way. They add items as do the other speech acts since, after all, they themselves are speech acts. As speech acts, they add either directives or commissives to the memory bank. But their main task in altering the context (of each of us) is to send a conversation in a certain direction and, in doing that, employing or invoking a different portion of the memory bank. Master speech acts have helpers such as modifiers that literally do what they say they do. They move the conversation to a different, but related, arena of the context if the speakers wish to move it in that direction.

So various speech acts can alter the context in many different ways. This fact might lead to the impression that context is completely malleable and so has little or no structure to it. It might be thought that it resembles chewing gum.

At this point it is important to remember that context possesses not only malleability but also stability. Logic acts as one stabilizing factor or agent. But stability is also found in the sciences, common empirical beliefs about the world (such as there are trees out there), and common ethical beliefs among other places. So, each of us possesses his own context that balances malleability with stability.

NOTES

1. John R. Searle, "A Taxonomy of Illocutionary Acts," *Expression and Meaning: Studies in the Theory of Speech Acts* (Cambridge, London, New York and Melbourne: Cambridge University Press, 1979), 16–17.

2. Richard V. Oulahan, "How the Rough Riders Got Their Name," *The Quarterly Review of Military History* (Vienna, VA: HistoryNet, LLC, Summer 2018), 32–37.

TEN
Putting it Together

SIMPLE EXAMPLES OF TOGETHERNESS

The purpose of most of the early chapters has been to identify many of the major clusters of presuppositions present in our memory when we are using language publicly or thinking by ourselves. Unfortunately, as we have seen, such a purpose can mislead. We language users don't create speech acts and speech activity (discourse) that display only a single layer of the context. Rather, we display many levels together, at the same time. The best way to show how this many-leveling process works is to present some real or real-like examples.

In this first example, two neighbors are discussing a mundane topic. Sara says: "Wow, it really rained a lot last night. I was awakened by the thunder and lightning. Then I found I couldn't go back to sleep." Tina replies: "Yes the rain, the thunder and the lightning were horrible. It awakened me, but eventually I went back to sleep." Sara replies: "You're lucky. I have to go to work soon and I'm tired even before I get started." Tina begins to end the conversation by saying: "Yes, I'm lucky, so right now I'm OK for today's workload. I hope you make it through the day." Sara replies: "Thanks, see you later."

Even this dull conversation is rich in context material. It starts with Sara's "Wow, it really rained a lot last night." This utterance serves at least three purposes. It informs Tina that she knows about the night's weather and what she thinks about it; and it serves as a master speech act to guide what is to be talked about as they meet briefly before they go to work. In the conversation that follows the master speech act, both Sara and Tina presuppose, among many other things, the following. It rained water, not molten lava, not rocks, not mud, etc. They also presuppose that gravity did its work in making the rain fall rather than rise into space. Further, they presuppose that the laws of physics apply to the building in which they live. It will not, they assume, turn into mud, change into another kind of building, etc. They also assume other crazy-sounding things.

As to Sarah's master speech act itself, she and Tina presuppose that both recognize what Sarah said as a master speech act. They both know that the job of such speech acts is to trigger other speech acts of a certain kind. No doubt they are not familiar with the concept, but they know how to pick up signals in speech that control later speech acts. In addition, they both presuppose that what Sara said and how Tina replied are words and phrases that they both understand. That is, they presuppose that certain meanings are in place.

Further, in being willing to issue certain assertive claims, both assume that each is speaking the truth (i.e., each is satisfying the sincerity condition). This assumption might be in question if Tina had been caught lying in the past. But unless such a reason is in play, the truth presupposition will be in play automatically as will certain normative rules and principles. Both Sara and Tina will assume that neither one will kill or steal from the other. Conversations of the Sara-and-Tina-type wouldn't take place if there were not a strong level of trust between them. There are still other assumptions present in their simple conversation. They both assume that one or the other won't die on that day, that they will see each other later, and that each will go to work that day. It is clear then that enough contextual material has been identified to show that we do, indeed, appeal to a host of presuppositions when we speak and do so all at once. The issuance is not brought off serially so that, it might seem, we would have time after each speech act issuance to ponder what all is presupposed. If that is the way it would go, we would never have time to process all the material found in the context. It is because we do most or all our presupposing "automatically," without thinking, that Sara and Tina can have their conversation and still get to work on time.

A second example differs in several respects from our first. It is a *New York Times* editorial titled, "A World Without OPEC" written by Joseph Nocera.[1] This example differs from the first one in that, to be understood, institutions need to be brought into Nocera's story line in a way that is not the case with Sara and Tina's conversation. To be sure, the ladies talk about their jobs. Let us assume that one works at IBM while the other is a lawyer at a small legal firm. And when they talk to one another, each knows where the other works. But their weather stories do not require that references be made to their places of work, so all that kind of information remains in their memory banks idling, as it were. This becomes clear when we imagine a third party who knows Sara and Tina but does not know where they work. Still, he can fully understand their comments about rain, lack of sleep, and going to work, and so on.

In contrast, a reader would not understand Nocera's reference to OPEC if she did not know what countries make up that entity, even if she knew that OPEC stands for Organization of Petroleum Exporting Countries. If she thought that the United States and Russia, two of the world's biggest oil producers, are its two biggest members, she would be dead wrong. To understand Nocera, it would not be necessary that she know the name of each and every member. It would be enough if she knew that Saudi Arabia, Iraq, Iran, and Venezuela are members. Another reader might also understand Nocera even though his membership list only includes Saudi Arabia, Kuwait, United Arab Emirates, and Ecuador. A third might have a somewhat different and longer members list. What they need—for us to say that they understand Nocera—is that their lists are correct and somewhat overlapping.

A more obvious difference between the weather story of Sara and Tina, and the Nocera editorial, is that there is only one listener in the former, whereas there are an unknown number of listeners (readers) in the latter. That creates a problem. It isn't just that many of his readers, the informed ones, have a different list of OPEC members in their memory banks. It is that some don't know enough about OPEC to understand what he is saying, yet he still wants to communicate with these know-nothings. The problem, however, is not serious for Nocera or, for that matter, for any author of an editorial, essay, book, or speech. He deals with these know-nothings by inserting information into his text to which others do not have access. He does this slyly so as not to offend them for their ignorance. In effect, Nocera educates them so that they can now understand his editorial and understand other writings on OPEC as well.

The reason the former know-nothings now can understand Nocera's editorial is not just that Nocera is a good educator. Their "wiring," concerned with other aspects of their memory bank, is already in place. They must have some sense of logic, know that oil comes from the ground, that it does not turn into drinking water, know about the effects of gravity, know that countries exist and do not change their place on the planet easily, know even that there is a real world all around them, that the world is not some invention of Bishop Berkeley, etc. In this sense, in educating his readers, Nocera is not putting into place a whole new wiring system. Rather, at most, he is adding a few wires to a very complicated memory system already in place. He is enriching their banks.

COMPLICATED TOGETHERNESS

If, indeed, this is the way things are, we can begin to understand how a still more complicated linguistic scenario might take place. In this scenario, two archeologists, Howard and Carter, are planning to do research on Tutankhamun's relatives. They know that only some information is available about the relatives and that with more effort more can be uncovered. They need to plan on how to proceed, develop a plan, estimate costs, write a proposal to help them obtain funding, and assign one or the other tasks that fit their skills (e.g., Carter will oversee the DNA work, while Howard will explore all the available records). Each will have other assignments as well.

Eventually, they get funding and so begin hiring specialists and general workers to get the work done. They also spend time buying (or leasing) equipment. Finally, they begin their research by traveling to Egypt.

Imagine, before all that gets started, a conversation between Howard and Carter. It begins with Howard suggesting that they discuss funding (i.e., by issuing a master speech act to that effect). And they do. They talk about how much money will be needed to keep their research going for three years or longer. A figure of five million dollars is bandied about to allow them to get started. They then start clicking off the names of foundations, institutes, and governments that might help. They settle on applying to foundations A and B, and the Egyptian government. Each of the institutions they discuss generates its own layers of presuppositions.

Having settled on their "target" institutions, they decide to discuss, in more detail, what sort of equipment and supplies they will need (after one of them issues a different master speech act). Both Howard and Carter contribute to the discussion. It soon becomes clear that the five-million-dollar figure will not be adequate. So now they talk about needing 5.7 million dollars. Now more layers or presuppositions get generated.

The issue of hiring people, both specialists and non-specialists, comes next (by one of them issuing a different master speech act). They both know many learned people and know as well to which agencies they can go to find a large number of hardworking and responsible general workers. Finally, with the issuance of still another master speech act, they outline the steps they will take to start their project. Several more layers of presuppositions get generated to facilitate that discussion.

What is new in this scenario is the phenomenon of context shifting. As noted already, shifters do not change the overall context. They bring about only modest changes. Thus, when Howard and Carter shift their brief social speech activity to issues related to funding for their proposed project, what might be called a small package of context items emerge as they carry on their conversation (by issuing new text material). In that package will be context items common to their discussion of what sources exist for their fund gathering, hiring, and organizing their research. These common or shared features of the context (and memory) will include all those discussed in earlier chapters under the heading of logic and logic-related matters. Other shared features include memories of many kinds of physical objects, such as buildings held in place by gravity that do not turn into gardens. In short, each new speech act introduced into a discussion carries with it a trail of presumptions, and these presumptions (expressed perhaps as new speech acts) carry with them their own trail of presumptions and, then, these new presumptions (expressed as speech acts) carry with them their own presumptions, on and on.

Other features of the funding talk that Howard and Carter engage in are specific to their talk. Here concepts such as "money," "interest rate," "re-payment schedules," "foundations," "governmental agencies," "grants," and "salaries" play a role. When the topic shifts to the equipment needed for their project, a new set of presuppositions come into play, along with the old ones about gravity, logic and the rest. The list would include "clothing," "digging tools," "flash lights," "food supplies," "tents" as well as more technical concepts such as "medicines" and "electronic equipment."

The shifting that goes on in their discussions with one another does not always have to rely on master speech acts. A discussion over coffee between two life-long friends can allow for unannounced subject shifting. For example, Joe, an old New York baseball fan, can say simply "New York won" while Ted, an old Boston fan, can counter with "But the Red Sox won too." Ted's response makes it appear that Joe's initial speech act about the New York team victory acts as a kind of master speech act that implies but does not explicitly say "Let's talk about baseball." But given what he said, Joe might have been urging Ted to talk about all kinds of New York City sports teams, most especially the New York Giant football team or the New York Knicks basketball team. Or he could have been urging Ted to talk about all kinds of news related to New York City. Given all this ambiguity, it is probably best to think of Joe's statement as one about a New York team and nothing more. Even so, this interpretation of what Joe is talking about would have all sorts of presuppositions embedded in it just like any other speech activity (i.e., text).

With or without the urging of master speech acts, it is pretty obvious that it is extremely easy to trigger context change. As we have seen already, such change can come from speech act to speech act as when Joe talks about the Yankees while Ted the Red Sox. And it can come by starting a new discussion ("Let's change the subject") or by shifting around within a more general topic as with the discussions between Howard and Carter over the issues surrounding the relatives of Tutankhamun.

DISAGREEMENTS AND TOGETHERNESS

For now, one other kind of scenario needs to be discussed before moving on. This scenario is concerned with disagreement. Imagine Abe, a social conservative, who claims that his main political agenda is protecting traditional family values. Not surprisingly, he is a pro-life person on the issue of abortion. In contrast, Liberty considers herself some sort of a left-winger. To say the least, they differ on a variety of social issues in addition to abortion.

Despite their differences, they share enough to be able to carry on "debates" about their differences. Here is a partial list of their shared presumptions:

1. For both, English is their native language.
2. They know enough logic to be able to distinguish good from bad arguments, at least some of the time.
3. They are aware of such facts as that the normal human gestation period is about nine months; and that viability is reached at about four and a half months.
4. They know that for more than a few generations, marriage was said to be between a man and a woman in their culture.
5. They know that modern birth control procedures are very reliable; and that some of them prevent the transfer of disease.
6. They know that the human population has increased to over 7 billion by the 21st century and continues to increase.
7. They know that modern health care and modern sanitation procedures have increased the life span of humans significantly.

They even find themselves agreeing on issues about which some might suppose they would disagree. They both, for instance, believe in supporting education via subsidies for students through college, and believe in raising teacher salaries. But their disagreements arise first and foremost over the issue of abortion. Abe is prone to say: "Abortion is the killing of an innocent unborn infant" while Liberty would never be caught saying anything like that. Abe is serious about abortion because his ontology includes souls. For him, even from the moment of their creation, unborn infants contain features that transform somewhat formless human creatures into fully developed infants that we can eventually hold in our arms. Not surprisingly, Abe's ontology includes God, and a host of other religious entities.

Liberty, in contrast, is content to include in her ontology such creatures as lions, dogs, chimps, dolphins, eagles, and snakes. But she is not content to live without a set of social and moral values that help her justify saying "a woman must be allowed to have an abortion." These values include commitments to the various rights that women have, and that women should be given equal social status to men. In short, she brings the whole force of feminist ideology to the abortion table just as Abe brings to that same table the whole force of his religious ideology.

Abe's and Liberty's argument about abortion dramatizes three points that have been implicit in this discussion of context (and memory) so far. The first is that context possesses a strong individual component. It doesn't matter for our purposes how each person gained this component. Biology certainly has a say in the matter, as does the social setting. What matters is that once these contextual features (memories) enter an individual person's psyche, they influence what that person says and means when she uses language, and the effect they have is different for another person because her psyche is constructed differently. Second, it appears that some components of the context (and memory bank), the ones we sometimes call ideological, have greater impact or weight on how language works (on the text) than others. Third, when we appreciate the impact that the context has on what we say (the text again) we can begin to appreciate why it is so difficult for people to change their views on a matter such as abortion. That is, when such a discussion takes place it is evident that it is not just a discussion on the topic of abortion. Rather, it is one where the topic is both on abortion and the ideological portion of the context that comes from the memory bank. The discussion, then, is not about one issue but is a whole package of issues.

TEXTING TO ONESELF

Before moving on, I would like to turn to a topic that I have avoided so far. Consider the following case. Andrea likes to write about her herself, her friends and about life in general. Her computer is full of reports of what she has experienced for ten years now. Her entries in her journal are of two kinds. They report the events of the day especially ones that affect her, but they also contain musings, sometimes several pages long, about her own and other people's cultures as well as local and world events. In short, she writes about whatever comes to her far-ranging mind. She shows what she writes to nobody, although she has a vague sense that she might eventually publish her journal.

There is a problem right off in assessing Andrea's writing in terms of what is the text and what is the surface portion of the context. One way of looking at it is that all her writing in her journal is the text. We sometimes think of a textbook in this way. The text is the whole book. But, in another way of looking at it, the text is merely the writing she is doing at this moment or in the next few minutes. The reason for looking at what the text and the context are in this way is that Andrea's writing of the moment, or of the next few moments, often presumes what she has written earlier in the day, the day before, etc. Even she is not likely to understand fully what she is writing right now if her memory of what she has said a while back has failed her.

Of course, she does not have the problem of memory loss the way our friends from Chapter 1 have with their casual discussion of Derek Jeter and his accomplishments. Since they have no record of what they said yesterday, that text and the context associated with it can be easily lost.

Andrea's computer keeps her from suffering a comparable loss. When she checks her computer she is able, with assurance, to make comments like "Now I see I am contradicting myself" or "I'm repeating myself." In short, her computer is her memory helper.

Aside from being better remembered because Andrea's work is recorded, what I am calling the surface context is the same for her the way it would for two or more people in conversation. It is much the same for the rest of the context. The context does much the same work when the speaker and hearer(s) are different as when, as with Andrea, the speaker and hearer are one. Like those in conversation, Andrea finds that logical, empirical, normative considerations do what they do. I will stretch the point now even though I will have more to say about the stretch later. If Andrea thinks about what she is going to write, but she has not yet written anything, the claim is that the context is there doing its work just as if she had put her thoughts on her computer screen.

There might be two possible changes in the context for Andrea as compared to our Jeter commentators. It might be thought that master speech acts would have no role to play in her conversation with herself. Recall that the standard setting for master speech acts to play a role is in a conversation where the speaker attempts to help the listener understand better the direction the speech activity is going to take. Speaking to herself, it would seem, this sort of help would not be needed. It is true that it might not be needed very much. But Andrea would probably find it useful to add titles and sub-titles to her work in order to keep her thoughts on course.

What more obviously wouldn't be shared with forms of speech, where the speaker and hearer is not the same person, is the sincerity condition. This condition is buried inside the speech act in order to commit the speaker to believing, for example, whatever assertive she utters. Failing to satisfy this condition gives the hearer an opportunity to criticize the speaker. However, since it makes sense for Andrea to say to herself "Do I really believe what I am saying?" it may be that the sincerity condition is still in place in her writing. No doubt, that condition plays a weaker role for her because she would not likely lie openly to herself the way she could to another person. She could only fail to meet the sincerity condition by fooling herself.

In general, then, there is no fundamental difference in the work done by the context when Andrea is talking to herself and when talking to others. The various features of the contexts are at work in both private and public settings.

But imagine the busy Andrea also writing a novel. Is she now operating under different conditions that might lead us to suppose that presuppositions of ordinary discourse are different from those in fiction? Let us suppose that Andrea is in the early stages of producing what she conceives to be her great masterpiece. When she works, Andrea likes to be away from the public so she travels to her mountain home and stays there for a week or so before returning. Then, after a few days of mingling with others, she returns to the mountain for another week or so. She takes this back and forth journey until she finishes her great work.

Because her novel is about urban living in America twenty or thirty years ago, she has read extensively about the sociology, psychology, and history of the period. She has filled her memory bank with lots of information that only a few others possess. She also has access to materials in her memory that she shares with others who have lived through the period or who happen to have read about it or have seen events of the period portrayed on mass media.

Andrea adds to her context memory in yet another way. The fictional characters she creates add more material to her memory bank. They do so gradually as she writes her third chapter, then the forth, and so on. By the end of her novel, her readers will understand why Abby, her main character, acted selfishly in the last chapter since she is acting in character in terms of what we know of her from Chapters 4 and 5.

All this new memory bank material is added to her larger memory even though Abby and her friends are fictional. This may not seem right, at least initially. But consider how we know about a real character in history. Napoleon is not known to us (directly) even though he was a real person. Of course, he was real; and we know that he was real. To us, his reality comes from history books and documents. The more we read about him the more his character becomes formed in our minds. It is the same with Andrea's fictional character, Abby. Like Napoleon, Abby is an empty shell until we read about what she does, what her expressed beliefs are and so forth. In terms of what gets placed in Andrea's (and later her reader's memory) it makes no difference whether she is real or not. As a result, Abby can inspire us or trigger a reaction of disgust just as (a description of the real) Napoleon does.

It is no different if the fiction is about a real person. William, another future author, presents us with an account of what a real person did but, then, also fictionalizes some of the things he did. Yes, Churchill led his country in a great war but, no, he did not say the things our author says he said on a certain day in December 1941. A historical novel like one William is writing is just as likely to contribute to the creation of a rich context memory as is other forms of fiction and history.

In this chapter, I discussed how the workings of the memory bank are the same, and how they are different in different kinds of settings. Some of the cases featured different kinds of assertives, others directives or evaluatives, and still others commissives. No matter what the combination, each example of speech activity always seems to carry with it layer upon layer of speech memory. That memory is not usually presented to us one at a time in serial order. Rather it is more like this. There is an appeal to layer #1, then #2, then back to #1, and from there to layer #4.

Finally, there may be a return to layer #2. On the face of it, it all looks very confusing.

NOTE

1. Joseph Nocera, "A World Without OPEC," *New York Times,* Tuesday, October 21, 2014, A25.

ELEVEN

Four Processes

LOOKING BACK

Start from the beginning and go from there to what has been learned since then. When a human adult uses language, that is, issues a text, it looks as if all his listeners need do in order to understand that text is to listen. The text, it seems, tells its own story, and tells it completely.[1] If there is a problem in understanding, the solution must be to look at it (hear it) again, this time more carefully, and the light of understanding will turn on.

A suspicion about such a bare view of language (use) might arise when Anne becomes angry because John continually misrepresents what she says. On one occasion she said, "I suspect it's going to rain." But John pays no attention to "suspect" and replies with "You said it's going to rain." In response, Anne says "You've taken what I said out of context."

That is easy. John who never thought about context before will, hopefully, come to realize that there is such a thing but, even if he does, he will likely feel that he is making only a minor adjustment to his overall picture of language as a bare, naked, phenomenon. For him, language (the text) still stands pretty much on its own. For him, talk about context is still window dressing. For Anne, context is about what was said a moment ago, or perhaps an hour ago, or even the previous day. But for John context (and the memory bank) is nothing more what we find on or near the surface of language use (see Chapter 1).

However, as it was made clear in later chapters, John just doesn't get it. Each person's memory bank stores layer upon layer of all kinds of "information" ready to be accessed when the occasion arises. When it does, the information moves to the level of the context where it is used immediately to make the text make (more) sense. The speed with which this rising happens suggests that what is found in the bank is not stored in linguistic form. If it were, the user's thinking (textual use) would stop in its tracks. If that were the form the context items took, matters would be even worse as the speaker went on to issue another speech act and then still another one, with each speech act trailing its own special contextual profile. Given the layer upon layer of the contextual items supporting each speech act, the support would have to be coded so that the processing could be done instantly. But the point here is not to identify how this processing takes place. That is the scientist's job. The point is that some sort of quick processing needs to be in place in order to avoid a massive textual log jam.

Sometimes, of course, we know we have to work our way through log jams. We stop the normal flow of a conversation, as Anne and John did, to get things straight. Hopefully, they can settle the issue as to whether Anne prefaced her comment about the weather with "suspect." Having done that, they can now continue their not very interesting discussion about the weather.

But even this simple, not very interesting weather example exhibits the four basic processes in which we engage when we use language, and thus become involved in appeals to context(s). These processes can be labeled creation, storage, retrieval, and inspection.

THE CREATION PROCESS

A speaker issues a master speech act followed by a series of five or six speech acts triggered by that act. Once presented, these speech acts are followed by another cluster of speech acts and then by another. Often, what breaks one cluster from another is that one speaker talks for a few moments and then a different speaker takes over. Once these two clusters of speech acts (texts) have been uttered, they drop into the memory banks of the speakers and listeners ready to be used to help them understand what is said later. The creative process here just is this process of adding to the memory bank and then, usually later, to the context.

It should be clear that what drops into the memory bank is largely, but not exclusively, text material. If I get angry with you because you have publicly insulted me, I not only remember what you said but also remember my negative response. There is a cognitive record in my brain about what you said, but also an attitudinal or an emotional one. What doesn't go down below to the bank are special kinds of memory (and context) material (e.g., having to do with logic, meaning, and grammar) that helped make it possible to understand the text as text initially. That stuff is in place before the text showed up in linguistic space. It was already in the bank ready to be used.

If this picture of creation is somewhat plausible, one can ask: but what happens to the "items" or materials when their roles in making texts (more) meaningful are over? Do they return to the bank or do they just disappear? It is difficult to know what to say here. Favoring the return option is the suggestion that the return, if that is what it is, reinforces the memory thus making it less likely that the portion of the bank that was used will be forgotten. More on that subject shortly.

The original text would be more complicated if what had been sent down to the memory bank were not only assertive speech acts, but a wide variety of directives (normative claims) as well. Now the context supporting the text would also contain normative speech acts and/or habits (tendencies) as well as speech acts and/or tendencies supporting assertive claims. Complicated or not, a creation process would have taken place. The memory bank would, then, in some small way, have been expanded by one text issuance after another. As each of us (seven billion or so) speaks, each modifies his very own memory bank. But, of course, our individual banks don't just take in what is new in textual form. They encode what they receive biologically. Just as banks these days convert cold cash and checks into electronic blips, so our neurological system engages its own form of conversion. Following the conversion, the process of storage begins.

A reminder is in order. The creative process is broader in scope than the conversational examples used above and in Chapter 1 would suggest. Creation takes place when we read a book, listen to the news on the radio, listen to a talk, lecture or sermon, and send emails back and forth. The creative process enriches our memory banks via many sources.[2]

STORAGE AND RETRIEVAL PROCESSES

The storage phase of the context story is largely peculiar to humans. Unlike banks (at least honest ones) the amount of credit you have stored away stays roughly the same. The credit you have drops when you pay for your new car and goes up when you add millions of dollars to your account as the result of your rich uncle's untimely death. In contrast, with the memory bank, what is stored gets lost without your doing anything. Much of that loss occurs on the level that we have called our surface memory. Conversations pack away everything we say but, quickly, memory fails us and so we forget details of what is said and, at times, forget the whole (superficial) conversation we engaged in with one of our idiot relatives. But even important memory items get lost in time. We forget the geometry lessons we learned (well) in high school, and the names of most of the Roman emperors we learned about in college. We may even forget how to engage in certain physical "know-how" activities such as how to shift manually from one gear to another in an automobile. The memory bank evidently is a leaky storage facility.

Even though it leaks, it still holds lots of information both of the knowing-that and the knowing-how variety that remains ready for employment in support of any new text that comes our way. So, imagine that a new text is issued by some speaker, and so the process of retrieval begins. The speaker talks, let us say, about everyone's favorite topic: viz., black holes. Now begins what looks like, given what we know about physics, a magic act. Somehow when something is said about black holes, the mind is capable of reaching into our own memory bank so as to alert to us about what we learned in our physics class years ago. At the same time, it ignores what we learned in the past about some distant relatives, cities, last year's news events, etc. This "magic" act places information about black holes in the context of the speech acts being issued at this moment and the next. The retrieval process fails when, as black hole speech acts are issued, nothing or very little is retrieved from the memory bank. It is successful when enough information is retrieved to make understood what otherwise would not be understood.

At this point, one must not forget that the retrieval process is not uniform in character. Some of the retrieval will be in the form of speech acts. So the hearer (and speaker as well) will understand what the speaker said about black holes when she runs through thoughts about black holes only when they are expressed as "Network" speech acts. But since texts concerning black holes, and other matters, are so complex, it would be impossible for any hearer to follow a series of black hole speech acts if she had to recite immediately one two, three, four, etc other speech acts. By the time the hearer has digested all the speech acts in the bank (and the context) that would supposedly have made it possible for her to understand the first speech act, the speaker would have uttered a second, third, fourth, etc., speech act. With each speech act uttered, the hearer would get further, and hopelessly, behind. At the end of an hour long lecture, our listener would likely be processing speech acts uttered fifty minutes ago.

On the context level, then, some other way than speech act reciting has to be employed to make it possible for the hearer to understand a lecture (a whole series of speech acts) on black holes. A much quicker way of retrieving needs to be in place. That way features the concept of knowing-how. Put aside the black hole example for a moment and go to the example of the player hitting a tennis ball. A new student is taught many rules. One is that if the ball coming at you appears to be landing near your feet, take a few steps back so it lands well in front of you. When you follow this rule, you are able to return the ball more forcefully since you are stepping forward into the ball. Another rule is that it is best to hit the ball with your wrist bent back (rather) than in parallel with the rest of your arm. Hitting the ball this way insures that it will go straight back rather than veer to one side (off the court) or the other. Once these, and many other, rules that were learned not only cognitively, but also through practice, they will disappear from consciousness. One no longer has to recite them each time the ball is hit. It may be, from time to time, that some recitation (i.e., inspection) is needed especially at first. Recitation may also be needed because one may have forgotten some detail (or the whole of) some rule. In time, when one becomes a professional in the game, no recitation is needed. One sees the ball land several feet in front of him, and he just hits it hard. At this point, the player's response seems to be instinctual.

Even so, although the rules are no longer recited, they are still in play (pardon the pun). But being in play now means they are presupposed. The player knows how to hit the ball (in accordance with the host of rules he has been taught), but he does not have to take the time to recite them. As a result, the player is able to do several things all in an instant.

Consider next a somewhat more intellectual matter. Susan does her calculating in her head. She does not like to use electronic or mechanical devices to do arithmetic. So, in time, her know-how in these matters is pretty sharp. She adds, subtracts, multiplies, and divides effortlessly. She knows all the rules found in her memory bank, but she has no need to appeal to them. The only time she appeals is when she is teaching arithmetic to her young daughter.

It is the same when Samantha discusses black holes with her fellow graduate students in physics. They discuss the properties of black holes without discussing such presuppositions about those very dark holes such as the speed of light, space-time, etc. Amongst the graduate students, discussing the number and nature of the presuppositions is a waste, unless, of course, Samantha and her friends are talking to or lecturing to undergraduates.

If such a "passive" account of context characterizes the role it plays in making many texts make sense, talking of moving from the memory bank to the context level is really a misleading metaphor. On this account, the portions of the bank that count as contexts don't really move at all. They stay in the memory bank. However, their status does change. In the bank, those features that make up the context are "lighted up," recognized, identified, remembered (its hard to know what words to use here since not really much is known about the physiology of these matters) and so are tied to the text. But, again, the items in the memory bank that count as context (when the text is uttered) do so because the text triggers (another metaphor) these items.

It's a little more complicated when discussing written texts since with writing the speaker (writer) is almost always separated from the hearer (reader). Initially, the text triggers the writer's context when he puts his thoughts on the computer. Then, as it were, the whole business goes to sleep until someone reads the text. Then when someone reads it, context materials (the alerted part of the memory bank) go into action. Although the context has to be shared for the reader to understand the author, sharing need not be perfect. What is required is that there should be extensive sharing.

One must also not forget that some context materials will not be in the form of knowing-how. Recall that so called surface contextual material can, and often does, get remembered as knowing-that. For example, I remember that when I got mad at you yesterday, I said "Go jump in the lake." But like the knowing-how context items, knowing-that items help make sense of the text both quickly and, to a certain extent, stealthily. Their quickness comes courtesy of how they are retrieved. They needn't be retrieved by a recitation of a whole speech act. "Go jump in the lake" as text does not have to be reproduced to help the text. All you have to remember is what you said, and you can do that without reproducing what you said.

THE INSPECTION PROCESS

The inspection process is a special form of the retrieval process. Unlike items that play a role as context for some text and, more often than not, keep their stealthy status, those items under inspection lose their stealthy status. They do so simply because they are brought all the way up to the textual level so they can be looked at clearly and carefully. It follows that they also lose their quickness. Rather than doing their work just at the moment a speech act is uttered and then resting again in the bank, they are looked at on the textual level possibly for a moment or two, or possibly for the length of an article, a manuscript; or still worse for years while scientists, ethicists, philosophers, theologians, etc. inspect what they are telling us.

We saw earlier why inspections might take place. From the beginning, it was clear that each individual has his own memory bank. To be sure, sharing is necessary if one individual has any hope of communicating with others. But even with extensive sharing, your memory bank will differ in some ways from mine. The differences can, and often do, signal serious conflict. So it makes sense to inspect your and my presuppositions to see if, in one way or another, we can determine whether they make sense.

But even when there is no conflict, inspection may be in order. After all, many of us realize that some of our assumptions have never been carefully thought through. We hold them, perhaps, because we were told to do so by our elders who, in turn, hold them because they were told to do so by their elders. Inspection might also be in order because we assume that what our memory banks contain got there at a time when our measuring instruments were cruder and less reliable than they are today.

There are, then, a variety of reasons why we should not always trust what is found in our memory banks.

In one way, the inspection process is not special. Like all other speech activity (extended texts), the process cannot operate without appealing to some parts of our memory banks. That is, not even the inspection process is context free. It might be tempting to think that it is. Thinking carelessly, one might suppose that the process could, or should, be comprehensive. We should be able to start from scratch by presupposing nothing. But the inspection process, like any process, needs context items in order to start, and then proceed to some conclusion. Further, even if we could carry out our inspection without appealing to contexts, the sheer scope of the process would be ridiculously beyond human capacity. The human brain simply cannot inspect all the widely varied kinds of speech acts in one fell swoop. The best we can do as inspectors is to focus on one, or a small cluster of presuppositions, and use some of the remaining presuppositions to further the inspection process.

Even if the inspection process is limited, it still is a daunting task to carry out. In part because inspecting any one item in our bank, whether it is of a concept or a full speech act, or set of speech acts, requires that at least some of its context also should be inspected. As we have just realized, the inspection is never just one that focuses on the item talked about in the text. Its roots also must be studied.

Although the inspection process is limited to studying one or a few items at a time, there is no limit as to what can be inspected or studied. Even the knowing-how features of our lives where the rules and principles have disappeared from consciousness as we develop habits of behavior and thinking can be inspected. Return to tennis once again. The good forehand habit has the player lay back his wrist when he hits the ball. He learned this way of hitting the ball from his coach by, at first, repeating this rule to himself time and time again and, actually hitting the ball in the correct fashion. Then, in time, this "unnatural" way of hitting the ball became natural. When this happened, the rule fades into the background. But it only disappears from consciousness while he is playing. While teaching the game to others he can, and does, call up the old rule all the way up to the text level. It's the same for all knowing-how linguistic and mental states. We can "forget" the rule without forgetting it.

If every item in the memory bank can be examined, are there some reasons for examining some items in the bank while ignoring others? One reason for initiating an inspection is if there is some superficial misunderstanding in a conversation or discussion. The inspection as to what was in fact said can take place quickly and easily simply by reminding one of the interlocutors what was said a moment ago.

A more serious inspection would be needed if the disagreement is over a scientific matter. There, the memories of all the discussants would be raised to the textual level so as to highlight and make clear what disagreements might exist. What would be raised and examined are not only the laws of science under question, but also their roots, that is, their own presuppositions. But not only would the inspection process involve looking at both the "flowers" and the roots carefully, it would also involve appealing to those parts of the memory bank such as other laws that relate to those laws that are in question. In science, there are also clear creative steps that (almost always) need to be taken. If the disagreement is over whether certain causal memories can be verified, then experiments yielding new results (memories) to verify or dis-verify the old memories might need to be done. With some luck and a good deal of time, this combined inspection process might lead to a "permanent" memory changes of those who understand these scientific matters.

Not surprisingly, the inspection process in normative arena is somewhat different. If some accepted an important ethical principle is up for inspection, data gathering might still be in order as in science. But the data would not be used in the same way. In science, the data would lead directly to modifying or replacing the old law with a new one. And that would be the end of it. In ethics the new ethical law arising from the data might now be only in place to act as a reason for altering the present ethics law or possibly replacing it. So there would be an additional step in the ethical inspection process before change might take place.

Another difference is that the procedures in these two domains are importantly not the same. In science the procedures for gathering data and even engaging in theory construction yield a considerable amount of shared agreement. Ethics, as has been commented on already, yields less agreement. To be sure, the common morality gives some agreement. But beyond that, Aristotelians, Kantians, utilitarians, anti-theory ethicists (e.g., casuists) are likely to generate a fair amount of disagreement. Even with these disagreements, there still will be some sharing. As in the sciences, logic will be the same, so will other parts of the memory bank such as the reality of physical objects (e.g., mountains, other people, and air).

RESULTS

With all the difficulties found in the inspection process, one cannot help but wonder about the quality of the results. It might seem that since the inspection process always rests on sets of assumptions, its results could not be anything but tentative, subjective and/or relative, and thus disappointing. Such wondering, I will argue, is simplistic.

Consider again the example of a text cited near the beginning of this chapter in which Anne says, "I think it is going to rain" and John takes her as saying "It is going to rain." Anne corrects John simply by referring to what is now in his memory bank. John's memory differs from Anne's.

At this point their little drama could come to a dead end. But suppose Anne remembers that she had her voice recorder working at the time, and so she replays their conversation. The voice recorder tells both of them that what Anne said was "I think it is going to rain." Of course, playing back what was said brings with it its own assumptions. But even so, if John is not a radical skeptic or paranoid, he will concede that he misremembered. Their superficial and brief and, no doubt incomplete, inspection of their memory banks has yielded an objective account of the text's and the memory bank's content.

To be sure, their accounts of their memory banks were not exhaustive. All they were concerned to report on was that portion of the bank/context that was in dispute. As to the rest, they adopted a "leave well enough alone" policy. If they foolishly thought that they had to inspect the all the assumptions built into their memories, and then inspect the assumptions of their inspection of the memories, the process would never end. The argument here is that the inspection process needs to end where serious puzzlement ends.

A criminal case yields similar results. Bill Jones is found dead in his bedroom. A gunshot to his head did him in. A gun lies next to him. It looks like suicide except there are no finger prints on the gun. There is no glove, handkerchief or anything of the sort at the crime scene to help explain the absence of fingerprints. So now it doesn't look like suicide. Immediate police work confirms that the gun was the weapon used to kill Bill, and that the only person who would gain financially from his death is his brother Adam. It is also clear that Adam was close enough to the scene of the crime to have killed Bill.

Understandably, the police form a hypothesis that Adam killed his brother. Working with the hypothesis, they discover that Adam bought the gun and ammunition a week ago. They also discover that Adam and Bill not only did not like one another, but that they had violent arguments recently about family matters. They even found a handkerchief in his laundry bag containing evidence of a gun being fired. So again, working on the hypothesis that Adam is guilty of the crime, they check the new evidence against gunpowder evidence at the crime scene. It all matches.

Notice that the police inspection does not question assumptions having to do with chemistry and physics. Indeed, they rely on the laws from each field to help tie the noose around John's neck. The inspection, instead, focuses exclusively on their hypothesis. They are keen on verifying it, or not.

One of the detectives, whose name of course is Holmes, becomes bit suspicious that the hypothesis they have all been using is flawed. His suspicions focus on the discovery that the argument between Bill and Adam was part of a larger argument involving their sister Joanne. In this argument, Joanne took Bill's side against the hated Adam. So Holmes begins testing the old hypothesis by developing an alternative hypothesis that Joanne may have had something to do with Bill's tragic end. With this new hypothesis in hand, new evidence begins to come in. Joanne was near the scene of the "crime" just as Adam was. Further Bill was suffering from depression. Even so, after Adam purchased the gun, he told both Joanne and Bill where he was keeping it.

To make a long story short, a suicide note in Bill's handwriting was found in Joanne's room. At that point, she confessed that she found her brother dead, took the handkerchief she found in her brother's shooting hand, took it to her hated brother's room so he would get blamed for the killing. So the inspection process was successful yielding, as it did, objective evidence of what happened.

In one part of the normativity domain, similar objectivity can be achieved but in a totally different way. Recall the common morality principles such as the no harm principle, the help principle, the truth telling principle, among others. When we think about them, it is comforting to realize that they are indeed shared in the memory banks of almost all people. All (or almost all) cultures articulate versions of these principles. But our comfort is disturbed somewhat when we recall how wrong the common belief that the world is flat turned out to be. But the common morality is common in a more comforting way. As pointed out already, when philosophers use whatever basic principle or principles animate their theory, they consider it a plus if their theories generate all or most of the common morality principles. So utilitarians suppose that their principle, when applied, will yield the-don't-harm principle, the help principle, and so on. Contractarians expect to get the same yield, as do Kantians, and virtue theorists. Anti-theorists often talk in terms of intuitions, but that talk also yields common morality principles. It is as if no theory or anti-theory approach dares to advocate an alternative account to the common morality.

This sort of proof, or justification process, is strange and perhaps not fully satisfying. But it would be stranger still if we ignored the judgments of professionals who have spent years studying the domain of ethics and then ignored the fact that their professional judgments which, of course, favor their own theory, all happen to be in agreement with respect to the common morality. It would be stranger still to look at the agreement arrived at by individual philosophers and label that agreement relativistic. The agreement comes from thinkers who have their own point of view, but that point of view doesn't seem to yield judgments about the common morality based on personal or group preferences.

Contrast all this agreement with the serious disagreements found in many other portions of the ethical domain. There is disagreement about war and whether to respond to it with an ethical theory of some sort, whether it be in some form of just war theory, pacifism, militarism, or even whether the response should be in the non-ethical tradition of realism. With respect to war, there are also issues of how it should be fought (e.g., with drones). In addition, there are disagreements about abortion, many business practices, dealing with refugees, financial inequality, the status of animals, whether to and how to deal with the environmental crisis, etc.

All, or perhaps most, of us have views about these issues packed away in our memory banks. We, of course, also have packed away memories that these views presuppose. The problem is that it seems we don't have a shared set of procedures or ways for settling these disagreements. Some thinkers fool themselves into thinking that there is procedural agreement. As I pointed out in Chapter 6, they do this by attacking those who disagree with them in a self serving way. Much of their criticism takes the form of position X is wrong because if fails to take account of features A, B, and C which features, curiously enough, their own theories possess. In short they beg the procedural question by saying to their opponents "You are wrong because you don't agree with my (correct) views."

I am not trying to canvass the whole of the inspection process of what is contained in memory bank's geography. Rather, the focus is on pointing to how varied this process is.

NOTES

1. Emma Borg, *Pursuing Meaning* (Oxford: Oxford University Press, 2012). Borg is a minimalist and, as such, tends to diminish the role of context in our language.
2. See the previous chapter.

TWELVE
Interpretation

EARLY STAGES OF INTERPRETATION

Most of the situations and cases discussed thus far have the speaker directly facing the hearer, thus making interchange relatively easy. If misunderstood, the speaker can usually explain himself to the satisfaction of the hearer.

Matters are a bit more complicated when the speaker faces a large audience. Still, some interchange is possible, and often sought during a question and answer period. The separation between the speaker and audience is not great; but it isn't insignificant either. It becomes greater, of course, if the speaker records his thoughts in a professional journal and then disappears back to his laboratory or office. He may be harder to find now, but an email or two can serve as intermittent conversation between the speaker and his listeners (readers).

But if an author dies suddenly, trying to understand (i.e., interpret) what she wrote becomes serious. Serious, but not overly so. If the author is a chemist and her article is published in a reputable professional journal, we can expect that her readers will have many shared memories about the profession and the work she has done. Meanings and scientific claims in what was in her memory bank will be much the same as that found in her readers' memory banks. However, matters become serious if our author died, not recently, but hundreds of years ago. With the ancient author separated from her readers (listeners) not only physically, but also in time, we can no longer be sure how much sharing exists between her and her readership. So it would be dangerous to read the author's text using the memory banks of contemporary readers.

The less dangerous course would be to match a text, written in 1500 AD, with a modified memory bank, one that, to the extent necessary, imitates the memory bank (and so the context) of the author's. But how can readers come to know that their imitating banks are really imitating? This is like asking Sherlock Holmes how he knows that the butler killed Lord Plumberry. Holmes knows that the son, daughter, the cook, etc. are not the killers since they were away when the crime was committed. He also has in his hands clusters of clues about the butler's thinking that all lead him to suppose that the butler is guilty.

In Holmesian fashion, anyone wanting to arrive at a most reasonable interpretation of an ancient document must look for behavioral and mental clues that the author left behind. No doubt the first place to start looking for such clues is in the text itself. Yes, the author wrote in ways that honor the principle of non-contradiction and the other basic logic rules. So, the reader's modified memory bank, now thought of as a (separate) bank annex, shares some things with the author. More sharing will likely be revealed by a careful re-reading of the text. The text may reveal, for instance, that what I am calling the common morality is being honored. As a result, it is becoming gradually clear that the annex memory bank is not going to be completely different from the reader's primary (i.e., contemporary) bank.

Still, an "inspector" looking at the text, most likely an historian, will focus on language usage. Let us suppose that the text makes frequent references to "family." Is this another example of sharing between an older culture and the present one? Suspicion might arise that it is not, since this concept seems to be used in the text quite often in an extended sense beyond referring to parents, grandparents, children, and other blood relatives. But, possibly, the inspector thinks, the author is speaking metaphorically. We too use "family" that way when we think of those with whom we work or play in sports as family. Thus, it is still an open question whether the uses in the text are metaphorical. Obviously, new clues need to be uncovered to settle the issue.

If it is later found that family members are allowed to marry and that family members are not necessarily buried near one another, the idea would naturally arise that the 1500 AD meaning is different from the 2019 AD meaning. But, let us suppose another clue surfaces. An additional reading of the text reveals that the 1500 AD people use the term "kin" close to how we use "family." So, for them, "family" means what we mean by "community," and "kin" means what we mean by "family."

It should be clearer now that the inspector who interprets our document works in much the same way as Sherlock Holmes does. Any kind of information found, either inside or outside the text, is open for consideration as evidence. That means that the inspector might very well need help from anthropologists, archeologists, psychologists, geologists, and others. Together, their goal would be to create an annex memory bank in each inspector that matches as much as possible the 1500 AD memory bank that the author had available to him when writing his text.

Note that this means that the consultant inspectors are not likely to be philosophers. To be sure, they might be, if the document in question is philosophical. In that case, some philosophers could be inspectors. However, let's say the document is an epic novel, a historical text, or something of the sort. Then, the inspectors will best not be philosophers, but are "detectives" in the field who have been doing their historical and anthropological work all of their lives. An analogy helps make this point. If one is interested in seeing how scientists work, it is best not to pay too much attention to what philosophers of science tell us about how scientific field work is done. Better to spend time watching how successful scientists work. Philosophers are not field inspectors or, at least, they are not likely good ones. Hence, it is the people in the field who do the hard work of filling in the details of the annex bank.

The reworking (aka the "creative" process) is not easy since, at least at the beginning, the investigators are likely to have only a partial picture of what they are looking for. Different documents from the past demand variations as to how they should be read. A document about war may demand a memory bank informed about military recruiting practices of the day, weather conditions during the days of battle, topographical information, population size, and so on. Another document concerned with palace politics, would surely not require battle related information gathered by military consultants. Whatever the information needed, once again, it should be clear that the skills for gathering such information are in the brains of the field workers. They are the specialists for gathering information about palace alliances, jealousies, etc.

Suppose now that the field workers have done extensive work, so they now know much more than before they first headed into the field. About our document, let us assume that it is partially fictional and was indeed published in 1500 AD, and that the inspectors have discovered other meaning changes besides "family." "Promise," for example, for those living in 1500 AD is used more to end discussions than to make firm commitments to do anything; and "getting revenge" is taken more seriously than it is today. For the people of 1500 AD, revenge would not be achieved unless two people on the "other side" are killed in exchange for the killing of one person on their side.

Researchers in our scenario have also made non-linguistic discoveries that seem relevant to understanding the text. Both the text and other documents of the age report that winters for many years were colder than normal. In addition, although the text tells of a major war taking place at that time, records show that only three small skirmishes took place back then. The text also tells of the general good health of the people when a major disease was actually decimating the population. Finally, the text reports that the "family's" leadership was receiving strong support from the people when in fact the family (i.e., the community) was close to rebellion.

Presumably these (and other) recently recovered "facts" were known to the author of our text (and thus were part of his memory bank) and, in uncovering these facts, our researchers are matching a certain part of their annex banks with the author's bank. Matching is important since our researchers are now more able to think like the author rather than typical 21st century readers. Still, in this sense of "understand," they have yet to achieve full understanding. What they understand better is that individual speech acts, and perhaps clusters of such acts, have changed in meaning over time. But what they might not understand yet is why the author's fictional story "misreports" the war situation, the weather and health conditions, and the status of the family leadership. Why, for example, is our author telling his readers that the family leadership was highly regarded when the real family was facing rebellion?

Some newly discovered facts could help explain this dissonance. Let us suppose the new facts are in the form of back-and-forth letters written by the author and a close friend. These letters reveal that the author was not just an author but also a friend and assistant to the family leader. That fact suggests a more macro interpretation of our document. It now becomes tempting to read it not as a descriptive account of what is going on in the family, but rather as a persuasive account aimed at making the family leader and his administration look good, thus to present a favorable legacy portrait for himself, his friends, and his administration.

At this point, the investigators' work is not over. All they have now is a plausible and relatively holistic account that helps them better understand their fictional document. More information might yet come their way to overturn their new analysis but even if the persuasive account turns out to be correct, the possibility of a second holistic theory being uncovered is not precluded. The author may have had more than one overall purpose or point in mind. In our case, in addition to a political document, it could also be seen as a religious one. The author might have dared to question the deities and their tendency to make the family suffer via cold winters, wars, droughts, and pestilences.

So the investigators (historians, anthropologists, theologians, etc.) have done some work beyond the new information they have gathered. To some extent, they are even thinking holistically. But, one wonders, if they should be engaged in still more holistic thinking and one might also wonder whether this more extensive holistic work should be handed over to a philosopher.

As suggested above, the answer is no! It is historians and other field workers who know best where the data they have gathered come from and how "valid" the data are. They also know best how to assess trends in human behavior and so are better positioned than philosophers are to combine the pieces of their area of expertise to form a pattern. But what, if anything, is there left for philosophers to do? At least this much. They can engage in analysis of such expressions as "understanding" and "meaning." These two concepts are, of course, important for anyone attempting to interpret a text. So are a related group of other concepts such as "explain," "comprehend," "realize," "imagine," and "identify with." Analysis of these and other like concepts capture, of course, part of what continental philosophers call the study of hermeneutics.[1]

LATER STAGES OF INTERPRETATION

By way of fitting some of these expressions into the process of interpreting a document, I will move away from our fictional family example to one found in science. It is a simple example that I have used before. An object is dropped from a great height. The physicist calculates the speed and exact location of the object on its way down. He does so by using certain instruments but also by appealing to certain laws of science. We say he understands what is happening by way of knowing those laws. We also say he explains or understands what happened when the object does, indeed, fall at the speed he said it would. But with falling objects, it makes no sense to go on to say that we identify with these objects as if we were continuing the process of understanding what is happening.

Nothing much changes when we turn to interpreting animal behavior. The laboratory rats have been underfed for several days. All have lost weight. When food is finally made available, all very actively pursue it. Again, laws are at play and we explain what happened by referring to those laws. We also say we understand what happened.

However, when we turn to some sophisticated animals, especially to human beings, things seem to be different. Some will say that they understand Jim when he lost his temper after a person he dislikes insulted his wife. But here there is no difference since our explanation of how Jim behaved is what we would expect of most anyone. His behavior is governed by a psychological law or law-like claim. However, the difference comes into play when Jim decides to join military service rather than keep working at the factory. Again, one can argue that the presence of a law still governs Jim's behavior. But the law is so weak statistically (involving several exceptions) that many would feel the law does not explain what he did. It also doesn't explain why Joe, who is very much like Jim and in a similar situation, chose not to enter the military. It seems now that we can come up with a satisfactory explanation of Jim's behavior only by following the trail of his thoughts. Having done that, we can now say that we open the possibility of (really) understanding what he did.

We can say this even if we find that Jim's thinking is flawed with respect to the evidence he cites for making his decisions. He believes, for instance, that the military pays more than his factory job when, in fact, it does not. Or, let us suppose, that his logic is flawed. His thinking involves making the simple error of reasoning from a grossly flawed sample. His best friend is much happier after joining the military than he was before. So, Jim concludes, from a sample of one, that he, too, would be happier in the military. Even with these flaws, the inspectors can say that they understand his thinking. They do so since, if they imaginatively adopt Jim's flaws, they realize that they would come to accept his conclusions.

With humans, then, there is no single way of arriving at understanding. At times inspectors will take the causal route as they do in understanding what other animals do. At other times they will take the reasoning route but there is no reason why they can't take both routes one at a time. They may want to do this in a situation where it is not clear that an author is actually moved by the reasoning process that he presents to his public. Thinking like Holmes, again, the inspectors should be open to receiving clues from any and all directions.

When the inspectors (the main historian scholar, anthropologists, archeologists, etc.) focus their attention not on causal forces that might be at play on our author of the 1500 AD document, but on how he was thinking, they will try to form a hypothesis about that thinking. After all, they can't directly access the dead author's mind. The only exception would be if, by sheer luck, the author explicitly revealed to them how the text should be interpreted. The author might say here that "the purpose of the text" is as follows or he might talk in terms of intentions. He might say "I intended to pass on to my readers the following moral (or religious, or whatever) message." The inspectors might still wonder whether the author was sincere in revealing his "purpose" or they might wonder if the author was fully aware of what he was doing. But still, the inspectors would be properly tempted to accept what the author said as a plausible interpretation of the text.

Of course, in many texts, the author does not explicitly tell us what he is doing. So again, the inspectors are forced to turn to generating hypotheses. Unless there is evidence that the author is crazy, they will assume his memory bank works, in part at least, in accordance with the principles of logic. Thus, they will gather evidence from the text and elsewhere to construct a pattern of the author's thinking that "makes sense."

Four final points need to be made about interpreting texts. First, the hypothesis that makes more sense after the inspectors have finished their work cannot ever be taken as final, since it is based, in part, on what they think went on in the author's head. Further, long and complicated texts are always open to reinterpretations especially if new facts emerge. Some of these reinterpretations could also come to light from personal reasons. A somewhat inexperienced researcher might insert some special items from his own memory bank into the annex bank. He might insert, for example, his own political ideology (e.g., Marxist or Conservative) into that bank. He might even be able to justify what he is doing by citing

Interpretation 93

"key" passages that suggest correspondence of his thoughts with those of the author. Still, it might seem to others that the new inspector is "stretching" things a bit.

More likely is that newer interpretations might appear because the new inspectors simply emphasize certain parts of the text to which the original inspectors paid less attention. For this and other reasons (cited above), we might say with justice that the interpretive process is never ending.

Second, the later process of interpretation could be more complicated than already indicated. Mention has already been made that it might be a mistake to develop one plausible interpretation and then to suppose that the investigators have finished their work. A second holistic interpretation, compatible with the first, might emerge later from the investigators' thinking. But beyond that, if the text is a long novel, it will likely be in order to offer interpretations of its main fictional characters. Such interpretations would be less holistic than the interpretation(s) of the novel itself and would be somewhat holistic if one or two characters are prominently present in the novel from the beginning to the end.

Third, the basic three steps in the interpretive process (reading the text, gathering data, and hypothesizing about what went on in the author's brain) should not be interpreted as taking place in serial order. They are explained that way in this chapter, but that explanation is in place simply for purposes of laying out the ideas in an orderly fashion. The process of arriving at an interpretation is more disorganized. Very likely it starts with a reading and rereading of the text. But after that, external data (other documents, weather conditions, social order in the author's society, etc.) are, if available, thrown into the mix. Once some data related to the text is gathered, very likely another re-reading is in order. That will be followed by more data gathering which might suggest an interpretation of the text that takes the author's thinking into account. With that interpretation in place, the investigators might return to the process of data gathering and/or still another re-reading of the texts. This back and forth process would continue throughout the interpretive process with which the investigators are engaged. Back to the data, back to the hypothesis (which now will likely be modified), back to more data gathering, back to a re-reading, etc. This back and forth process, whatever you call it (e.g., hermeneutic circle[2] which perhaps is not a circle but a spiral gradually moving upward, or a wave going up and down, and then up and down again moving from left to right) is common to all research. There is nothing special about the process. It's the way things are done in text interpretation, physics, chemistry, sociology, history, political science, you name it.

Fourth, more needs to be said about getting into the mind of the author. Ideally what the inspectors want can be expressed in more than one way. They want to "get into his boots," to think and have feelings like him, or even to identify with him cognitively and emotionally. All this sounds dangerous especially if the author is a monster and is expressing monstrous ideas (e.g., as with Hitler's *Mein Kampf*). But even if he isn't, you may not want to get so involved in your study of the author's text that your sense of who you are changes significantly. But, the question is, how can you be you after the mind of another person has been sucked into yours?

Given the distinction between your memory bank (and context) profile and the profile of the memory annex, the answer is simple. When you step into the author's boots you don't do that with the profile of your very own memory bank. You do it with the annex bank of the author but created by you. It's like you are on stage playing the role of Iago. But when you leave the stage you no longer are Iago. You are the actor who has a wife, children, and friends. Off stage, you are who you are before your stage appearance and after.

In truth, the inspectors of a text might change and so develop new memory banks of their own as the result of their research. They might be influenced, for example, by the exemplary behavior of the text's main hero. That sort of change might or might not happen. If it does happen we might now say "That's nice, now the inspectors are better people than they were before." But if that sort of change happens, the text's interpretation does not change.

To understand this point better it is helpful to return to a distinction that J. L. Austin made in *How to Do Things with Words* between illocutionary and perlocutionary acts.[3] A successful or happy illocutionary act represents a communicative accomplishment. The speaker (or writer) says something and the hearer (reader) understands what was said. An illocutionary performance is a linguistic act. A perlocutionary act is not a linguistic act. It has to do with the causal effect the illocutionary act has on the hearer (reader). Putting aside Austin's jargon, what we have is a happy linguistic act which, when completed, causes the hearer (reader) to respond in a certain way (e.g., by being shocked, obedient, made happy).

In making his famous distinction, Austin was concerned with how speech acts work. But his distinction can be applied to texts of almost any length (i.e., to speech activity). When it is, we can make more sense in saying that the investigators of a text have completed their work when they have successfully (more or less) come to understand what the text is communicating. As part of this investigation, the inspectors will recognize and respond to the normative claims stated or implied in the text. Their recognition and response will involve, as noted already, getting the inspectors to think and feel like the author.

They may want to investigate, as well, the causal effects the text brings about, but that is a separate task from interpreting the text. As a separate task, it requires a separate sociological, psychological, etc., investigation as to how readers of the text actually responded to that text. It is a legitimate investigation but not one of interpretation.

NOTES

1. Rudolf A. Makkreel, *Orientation and Judgment in Hermeneutics* (Chicago: University of Chicago Press, 2015).

2. C. Mantzavinow, "Hermeneutics," Stanford Encyclopedia of Philosophy. First published June 2016. See especially Section 2 titled "The Hermeneutic Circle."

3. J. L. Austin, *How to Do Things with Words*, second edition by J. O. Urmson and Marina Sbisà (Cambridge, MA: Harvard University Press, 1975). For our purposes, Chapters (Lectures) VIII and IX are the important ones.

THIRTEEN
Closing Thoughts

ANALYTIC/SYNTHETIC DISTINCTION

It is instructive to compare the "geography" lessons of this work with the lessons of W. V. Quine's works. Both lessons tell us about the importance of context in understanding language and thought, and the lessons coincide to a significant degree. But, as we will see, there are some major differences.

As is well known, one of Quine's favorite metaphors in understanding context is the web (or field) of belief. It is a metaphor that he applies largely to the discourse of science. The following passage tells us much of what Quine says about these matters.

> The totality of our so-called knowledge or beliefs from the most causal matters of geography and history to the profoundest laws of atomic physics of even of pure mathematics and logic, is a man-made fabric which impinges on experience only along the edges. Or, to change the figure, total science is like a field of force whose boundary conditions are experience. A conflict with experience at the periphery occasions readjustment in the interior of the field. Truth values have to be redistributed over some of our statements. Re-evaluation of some statements entails re-evaluation of others, because of their logical interconnections—the logical laws being in turn simply certain further statements of the system, further elements of the field. Having re-evaluated one statement we must re-evaluate some others, whether they be statements logically connected with the first or whether they be the statements of logical connections themselves. But the total field is so underdetermined by the boundary conditions, experience, that there is much latitude of choice as to what statements to re-evaluate in the light of any single contrary experience. No particular experiences are linked with any particular statements in the interior of the field except indi-

97

rectly through considerations of equilibrium affecting the field as a whole.¹

Quine rarely uses "context" in his writing, but his holistic account is quite at home with this concept. For Quine, no individual empirical claim stands on its own. It needs support from claims found here and there in the field (web). All claims, statements, speech acts, etc. help form in some small way the contextual bed. So far, so good. What he says and what our geography lesson tells us coincide. Also, so far, one difference is that the scope of Quine's thinking largely, but not exclusively, reflects the narrowness of thinking found early on in the analytic tradition. It is the narrowness that focuses on assertives (i.e., scientific and logical claims that are either true or false).

But there is another difference present. Quine attacks what he calls two dogmas that are found in what looks like his own tradition. These dogmas are reductionism and the analytic/synthetic distinction. It is the latter dogma that is of concern here. For Quine the criterion for identifying an analytic claim is certainty. The positivists, among others, say that an analytic claim, when true, is true come what may.

> . . . it is misleading to speak of the empirical content of an individual statement—especially if it be a statement at all remote from the experimental periphery of the field. Furthermore, it becomes folly to seek a boundary between synthetic statements, which hold contingently on experience, and analytic statements which hold come what may. Any statement can be held true come what may, if we make drastic enough adjustments elsewhere in the system. Even a statement very close to the periphery can be held true in the face of recalcitrant experience by pleading hallucination or by amending certain statements of the kind called logical laws. Conversely, by the same token, no statement is immune to revision.²

For Quine, an analytic statement would be true, we might say, in all possible worlds and for all times. It would be true because it cannot ever be false. If we found examples of analytic claims, they would contrast with synthetic claims that are probabilistic. But for Quine, there is no analytic/synthetic distinction simply because no claim meets the certainty standard of analyticity. That standard held by many analytic philosophers early on in the 20th century defines what analyticity means.

One way to attack Quine is to attack the dogmatic standard or criterion of certainty by replacing it with one that I will call the multiple assessment criterion. Appealing to this criterion, the distinction between analytic and synthetic claims reappears if the way each kind of claim is assessed is radically different. Grice and Strawson give us a perfectly good example of this difference.³ They ask us to imagine a three-year-old child described as being able to understand and explain Bertrand Russell's Theory of Types. Responses to such a claim would likely be something like "You can't be serious" or "That's impossible." If, in addition, the speaker says that he is speaking literally when he says "This three-year-old child is an adult" he might receive the same kinds of responses. If so, and if the first "impossible" is synthetic it would be tempting to suppose that the second is synthetic as well. So far, then, the analytic/synthetic distinction has disappeared again since both claims about the child receive the same response.

But if we push the assessment further, differences appear. We would naturally assess the Theory of Types claim by calling in the three-year-old and asking him to recite and defend the theory. But we would respond differently to the claim that the three-year-old is an adult. This time we wouldn't call the child in so as to inspect his mind and/or body. Rather, we would inspect the claims made by the speaker. We would suggest to him that he doesn't really know the meaning of "adult" as it is used by the vast majority of speakers. That is, we now would appeal to the shared meaning rules in our memory bank and not to principles dealing with observations of the child.

So the initial use of "impossible" in both cases hides from us a fundamental difference as to how we assess each claim. The Theory of Types claim is synthetic; the adult claim is analytic. The underlying reason for this difference, once again, is that each example appeals to different parts of the memory bank.

A similar point can be expressed in a different way. Quine's web of belief is not totally insensitive to differences inside the web of science with which he is concerned. He would, I am sure, claim that there are differences between theoretical claims and ones concerned with the laws of science. But the web metaphor encourages thinkers to play down these differences. So it doesn't seem totally unnatural to think that all claims on the web are on a dimension of being more or less alike. That, in turn, leads Quine and his followers to attack any claim that a class of sentences has the feature of a single criterion of absoluteness. Not only are all empirical claims probable, but so are definitions (they too can change), mathematical claims (they too can change), etc. They all belong someplace on the web.

If instead of thinking mainly of claims in the domain of science, one spends more time associating with claims outside of that domain, it is less tempting to believe that the lack of certainty explains all. Our memory banks are not best characterized as webs whose parts gradually (smoothly) blend into one another. Instead, one now sees memories (discourse) in this domain (e.g., law, ethics) and then that domain assessed, each in its own way. One is also tempted, as we have seen, not to look at all memories within a domain as being alike. We saw in the empirical domain, for example, how the memory records certain basic "truths" that are known without the aid of any of the sciences. Thus we have in our brains beliefs such as that the persons with whom we are having dinner are real, and that the trees and the mountains just behind us are also real. But we hold other empirical beliefs or "truths" because the chemist has run certain tests that tell us that the physical object in my hand has such and such metal fragments in it. Similarly, in ethics, we talk about one or more of the common morality principles and yet also talk about those principles or rules that do not seem to be so common. We are encouraged to talk about these, and other things, because the continental model being presented here encourages us to think of the memory bank as made up of many quite different domains.

Some of the most important domains to have appeared include the following:

1. The investigative domain. There are two basic speech acts featured in this domain: directives (that tell us how to think) and evaluatives (that rank almost anything on some scale or other). Assertives are also present here to provide reasons for any judgments that might need to be made. The list of domains and sub-domains itself is not what is important. What is, is the notion that the investigative domain contains certain laws and principles (e.g., law of non-contradiction, universalizability, the various laws of logic, induction). The emphasis here is on how best to deal with new problems and also how to assess accepted principles and laws.
2. Empirical domain. This domain divides in two. First, there is the pre-scientific (aka basic) sub-domain of such things as mountains, the earth, other people, and personal and inter-personal consciousness. All of us know about these things before science as we know it got going, and before we were told about science in school. Second, there are the various sciences each with its batch of laws and theories. Memories of scientific matters are not necessarily shared by all. These memories are aided by memory helpers such as books, videos, etc. The featured speech acts in this domain are, of course, assertives.
3. Social domain (empirical version). In spite of the objective nature of the social sciences, disagreement in this domain is quite common.
4. Ethical domain. This domain divides in two as does the empirical. First, there is the common morality composed of such principles as "Don't harm others," "Act justly," and "Do not deceive." Second, there is a wide variety of rules and principles shared by many (but not all) concerned with burial practices, assisted suicide, sexual practices, war, etc. Applying the principles of the investigative domain to the disagreements found here does not seem to resolve disagreements easily. This is so, in part, because there is no agreed upon set of investigative tools for settling disagreements the way there is in the scientific sub-domain. There are four basic speech

acts in this domain: directives (that tell us what to do), commissives (that create duties), evaluatives (that rank almost anything), and declarations (that assign duties and grant rights). Assertives are also present to help the process of reason giving that is inherent in this domain.
5. Prudential domain. Superficially, this domain is similar to the ethical. But this domain does not emphasize commissives and does not apply the universalizability principle. However, like ethics, it employs (empirical) data in order to calculate what consequences follow from choosing one option over another. The featured basic speech acts here are assertives, directives, possibly evaluatives, and declarations.
6. Political advocacy domain. The investigative domain does not seem to help this domain much at all. Disagreements here are rampant.

The list as I have presented it is also not important. There are other clusters of speech activity that are domains or domain-like that I have not seriously considered in this study. Aesthetics, thought of as a domain (a continent) or as domain-like (an island), is a good example. Religious talk and thought are another. My only excuse for not including them in the list is that I personally have never carefully explored these continents or islands. If these domains are like continents, and so are distinct from one another, they do not form a gigantic single web as Quine suggests they do. Relevant to all this discussion is another distinction.

FACT/VALUE (NORMATIVE) DISTINCTION

The fact/value story is pretty much the same as is the analytic/synthetic distinction story. The maligners say the distinction makes sense only if they can find large groups of "pure" facts and match these facts against a large group of pure value claims (i.e., directives, evaluatives, commissives, and declarations). Finding purity neither on the factual nor value side tempts many writers simply to deny that a distinction exists.

But the problem rests with the criterion for identifying a fact or a "value." If instead of looking for purity one looks for how we identify facts and values, one realizes that there are separate sets of criteria present. Factual criteria have to do largely with how observations are made while value criteria have to do with reason giving. If, as the example used in an earlier chapter tells us, the judge has issued a master speech act to the witness that she is to just "report the facts," and if she is a good witness, she will just tell the court what she clearly saw and heard. She would not add any normative comments to the story she tells. The same would be true of the hospital nurse who just reports the temperature and blood pressure of the patient. In doing that, and no more, she is appealing to a portion of her memory bank that tells her how a proper description is to be made. She implicitly makes the fact/value distinction by leaving the significance of her measurements to be determined by the doctor.

By discussing these two controversial distinctions (analytic/synthetic and fact/value), I am really doing nothing unusual. Language, whether used directly for purposes of communication or for private mental thought, is built to make distinctions. Thought, using language, and language itself are distinction-making machines. So it is not surprising that distinctions are being made all the way down (or up). Within each "continent," rivers can divide so as to make distinctions between common and not so common morality, and basic facts and scientific ones. And, for example, within the science, there are divisions between biology, chemistry, and physics. Further each science can be divided, as physics can between astronomy, the study of everyday objects, and the study of subatomic entities. All this is obvious.

But we need to be reminded that not all people make all of these distinctions. Some people's memory banks are curtailed by the limited experiences they have had. So, yes, their memory banks are full of ordinary facts, but are almost devoid of facts, values, insights, etc. that others possess. What makes the distinctions give forth with the impression that they are real is the concept of sharing.

Finally, there is one other feature of our memory banks that requires more attention. In the previous chapter, I introduced the notion of an annex to our memory banks. Using this concept, we can interpret a document, and some of the major figures portrayed in it without changing the "core" of our memory bank. Instead, we develop an annex that we access whenever we wish to "get into the boots" of another. We can access the annex and think like the author, but then leave the annex so we can resume our normal life with our "core" memories.

Well now, suppose that our text inspector moves on in order to interpret a second document. If he follows the pattern of his earlier work, he will develop a second annex to, again, become able to "get into the boots" of his second author. What if, after several years of marriage, our inspector finally becomes interested in interpreting his wife's behavior, attitudes, and feelings. Will he need a third annex? Of course he will. And he might need more annexes if he is keen on understanding other important people in his life.

Annexes might be needed as well when dealing with social matters. Our inquisitive inspector might be concerned to interpret the company he works for. Or, if he is politically inclined, he might wish to give an interpretation of the Supreme Court, a regiment he belonged to a few years back, or the church to which he belongs.

It's an open question whether our inspector could go on to create another annex to separate his thinking about physics, his specialty, from other parts of his life. Of course, there would be no issue pertaining to "getting into the boots" of physics laws and theories, But it might make sense to have a physics annex available since the thinking required to deal with that field is significantly different from most everyday kind of thinking.

It may be an open question whether it makes sense to think that the creation of annexes for studying the thoughts of physicists when they are thinking about physics. But what is not an open question is that some people can develop one, two, or three annexes in their memory banks. There seems to be a limit as to the number of annexes people can create because the process is so time consuming and difficult. But there are exceptions. A detective would be an example. So would a psychiatrist. Over time, the detective deals with a series of cases where he gets "into the boots" of one criminal mind, and then moves onto another one. Similarly, a psychiatrist, if he is a good one, "gets into the boots" of one patient after another. In effect, in order to get their work done, both the detective and the psychiatrist create temporary memory annexes. They become professional annex builders. Some professionals are precluded from building more than one or two annexes by the nature of their work. The annex memory bank of an ancient-text investigator could last for several years, and even could be a life-long project. Yet, on a more temporary basis, certain members of a theater group and their audiences could, and do, create many annexes as they come to identify with one and then another of the key characters in plays they perform.

But for many (some?) people who live day-to-day, their memory banks do not contain annexes of any kind. They worry daily about getting food on the table for their kids and themselves, paying the rent, etc., not about interpreting anyone's character.

FINAL GEOGRAPHIC PICTURE

Finally, then, our three-fold basic concepts in this study, that is, texts, contexts, and memory banks, look as follows. In this final look, I start with the memory bank, move to texts, and then deal with contexts.

Memory banks. As we have just seen, these banks can, but need not, grow annexes. Relative to the headquarters bank, annexes are few in number and probably small in size. Their main purpose is to hold information about special individuals and groups whose thinking needs to be explored. A much larger function of memory banks is to hold in the bank (i.e., memorize) all kinds of meanings, general information, general attitudes, etc. In this sense of "to hold in the bank," we are concerned with what I have been calling the bank's abstract content.

In addition to this content, there is strictly personal content material in the bank. Only the speaker (or writer) knows what she thought yesterday when one of her friends turned her down when she needed help.

What is in each person's bank is, in part, unique. Your memories just can't be like mine. As our experiences in life vary, so do the "things" found in our memory banks. Some of us have rather empty (and probably boring) banks. Others of us are lucky and so have enriched ones. Memory banks may have been created equal but, as life unfolds, these banks are clearly not equal.

Text. The text can be composed of just one speech act or thousands. Its range can vary as well. When Melissa talks to herself or when she secretly writes about her activities of the day, the range of her text (language use or thoughts) is one (day). When she broadcasts her views on Twitter, the range may be in the hundreds of thousands. The content of the texts is as varied as that of a person's memory bank.

Context. The context is dependent on the text to determine what items from the memory bank are needed to help us all understand what the text means or to help us understand the text better. It is also dependent on the memory bank to give it information, etc. so it can be helpful. By itself the context is empty headed.

NOTES

1. W. V. Quine, "Two Dogmas of Empiricism," in *Analytic Philosophy, an Anthology*, 2nd Edition. Edited by A. P. Martinich and David Sosa (Chichester, West Sussex: Wiley-Blackwell, 2011), 528. Originally published in *The Philosophical Review*, 60 (1951), 20-43. See also *Web of Belief*, second edition, by Quine and J. S. Ullian (New York: Random House, 1970).

2. Ibid., 528.

3. H. P. Grice and P. F. Strawson, "In Defense of a Dogma," *Analytic Philosophy: An Anthology*, Second Edition. Edited by A. P. Martinich and David Sosa (Chichester, West Sussex: Wiley-Blackwell, 2012), 536–537. First published in *Philosophical Review*, 65, 1956, 141–158.

Bibliography

Austin, J. L. *How to Do Things with Words,* Second Edition. Edited by J. O. Urmson and Marina Sbisà. Cambridge, MA: Harvard University Press, 1962, 1975.
Bach, Kent, and Robert M. Harnish. *Linguistic Communications and Speech Acts.* Cambridge, MA, and London: The MIT Press, 1979. Second Printing, 1980.
Beauchamp, Tom L. "A Defense of the Common Morality," *Kennedy Institute of Ethics Journal,* Vol. 13, No. 3, 2003.
Borg, Emma. *Pursuing Meaning.* Oxford: Oxford University Press, 2012.
Coleman, Stephen. *Military Ethics: An Introduction with Case Studies.* New York and Oxford: Oxford University Press, 2013.
Copi, Irving M., Carl Cohen, and Kenneth McMahon. *Introduction to Logic,* 14th Edition. Boston: Prentice Hall, 2011.
Draper, Kai. *War and Individual Rights.* New York: Oxford University Press, 2016.
Fotion, N. "Master Speech Acts," *The Philosophical Quarterly,* Vol. 21, No. 84, July 1971.
Fotion, N. "Speech Activity and Language Use," *Philosophia,* Vol. 8, No. 4, October 1979.
Fotion, N. *Theory vs. Anti-Theory in Ethics: A Misconceived Conflict.* New York: Oxford University Press, 2014.
Fotion, N., and D. Seanor. "Basic Acts: Is Five the Magic Number?" *Philosophical Inquiry,* Vol. XXV, Winter-Spring 2003, No. 1–2.
Freeman, R. Edward. *Strategic Management: A Stakeholder Approach.* Boston: Pitman, 1984.
Grice, H. P., and P. F. Strawson. "In Defense of a Dogma," *Analytic Philosophy: An Anthology.* Second Edition, edited by A. P. Martinich and David Sosa. Chichester, West Sussex: Wiley-Blackwell, 2012.
Habermas, Jürgen. "Discourse Ethics: Notes on a Program of Philosophical Justification," *Moral Discourse and Practice: Some Philosophical Approaches.* Edited by Stephen Darwall, Allan Gibbard, and Peter Railton. New York, Oxford: Oxford University Press, 1997.
Hare, R. M. *Moral Thinking: Its Levels, Method and Point.* Oxford: Clarendon Press, 1981.
Kaplan, David. "Demonstratives," in *Themes from Kaplan,* J. Almog, et al., Editors. New York and Oxford: Oxford University Press, 1989.
Kepner, Tyler. "In Enemy Territory: A Farewell Comes with a Warm Embrace," *The New York Times,* September 29, 2014, D1, D2.
Lee, Steven P. *Ethics and War: An Introduction.* Cambridge: Cambridge University Press, 2012.
Lewis, David. *Papers in Philosophical Logic.* Cambridge: Cambridge University Press, 1998.
Liao, S. Matthew, Editor. *Moral Brains: The Neuroscience of Morality.* New York: Oxford University Press, 2016.
Makkreel, Rudolf A. *Orientation and Judgment in Hermeneutics.* Chicago: University of Chicago Press, 2015.
Mantzavinow, C. *Hermeneutics.* Stanford Encyclopedia of Philosophy, June 2016.
Montague, Richard. *Formal Philosophy: Selected Papers of Richard Montague,* R. Thomason, Editor. New Haven: Yale University Press, 1974.
Nocera, Joseph. "A World Without OPEC." *New York Times.* Tuesday, October 21, 2014, A25.

Oulahan, Richard V. "How the Rough Riders Got Their Name." *The Quarterly Review of Military History*. Vienna, VA: HistoryNet, LLC, Summer 2018.

Parfit, Derek. *On What Matters: Volume One*. Oxford: Oxford University Press, 2011.

Quine, W. V. "Two Dogmas of Empiricism." *Analytic Philosophy, an Anthology: Second Edition*. Edited by A. P. Martinich and David Sosa. Chichester, West Sussex: Wiley-Blackwell, 2011. Originally published in *The Philosophical Review*, 60, 1951.

Quine, W. V., and J. S. Ullian. *The Web of Belief: Second Edition*. New York: Random House, 1970.

Ross, W. D. *The Right and the Good*. Oxford: Clarendon Press, 1930.

Scanlon, T. M. *What We Owe to Each Other*. Cambridge, MA: The Belknap Press of Harvard University Press, 1998.

Searle, John R. "A Taxonomy of Illocutionary Acts." *Expression and Meaning: Studies in the Theory of Speech Acts*. Cambridge: Cambridge University Press, 1979.

Searle, John R. *Intentionality: An Essay in the Philosophy of Mind*. Cambridge: Cambridge University Press, 1983.

Shaw, William H. *Utilitarianism and the Ethics of War*. Abingdon Oxon OX: Routledge, 2016.

Stalnaker, Robert. *Context*. Oxford: Oxford University Press, 2014.

Tuchman, Barbara W. *The Guns of August*. New York: The Macmillan Company, 1962.

Walzer, Michael. *Just and Unjust Wars: A Moral Argument with Historical Illustrations*, 4th Edition. New York: Basic Books, 2006.

Index

Almog, J., viiin1
analytic/synthetic distinction, 98, 99, 101, 102
Austin, J. L., vii, ixn2, 48n8, 94, 95n3

Bach, Kent, ixn2
Background, The, 28, 29, 30
basic beliefs. *See* common beliefs
Beauchamp, Tom L., 25n1
Borg, Emma, 85n1
Brooke, Rupert, 17, 18

Coleman, Stephen, 25n7
common beliefs (empirical), 27, 28, 30, 53
common ground (memory). *See* memory bank
communication, 42, 43, 44, 46, 51, 57, 102
conditions of satisfaction, 28, 29, 33n2
context, linguistic, 1, 3, 7, 9, 10, 17, 18, 29, 41, 47, 50, 51, 60, 61, 65, 66, 68, 69, 70, 72, 75, 76, 77, 79, 80, 81, 82, 83, 97, 103, 104; changing the, 2, 3, 5, 6, 7, 68, 69; out of, 6, 75, 83; surface of, 5, 6, 7, 9
communicative duties of hearers, 45
communicative duties of speakers, 43, 44, 45
community rights and the hearer's community duties, 46
contractualism, 38, 84
Copi, Irving M., 6, 8n8

direction of fit, 33n2
discourse. *See* speech activity
Draper, Kai, 25n9

empirical claims. *See* assertives

ethical thinking, 4, 21, 22, 23, 37, 38, 39, 42, 53, 62, 70, 82, 84, 100; common (morality), 21, 22, 24, 53, 84, 88; ethical principles, 22, 23, 24, 37, 52, 82, 84; exceptions to, 23, 37; mid level, 24; super, 22, 23, 24

fact/value distinction, 101, 102
Fotion, Nicholas, 8n2, 8n3, 8n4, 8n6, 25n4, 39n1
Freeman, R. Edward, 25n6

Garbo, 42, 43, 44
grammar, 4, 13
Grice, H. P., 98, 104n3

Habermas, Jürgen, 41, 42, 43, 44, 46, 47, 48n1, 48n2, 51
habits. *See* knowing how
Hare, R. M., 48n4, 48n9
Harnish, Robert M., ixn2
hermeneutics. *See* interpretation

illocutionary acts, 94
intentional states, 28, 29, 33n2
interpretation, 88, 90, 91, 92, 93, 102

Jeter, Derek, 1, 3
Just War Theory, 24, 84

Kantian thinking, 24, 38, 39, 82, 84
Kaplan, David, viiin1
Kepner, Tyler, 7n1
knowing how, 13, 29, 30, 36, 38, 45, 50, 51, 53, 66, 77, 78, 79, 80, 81
knowing that, 35, 38, 45, 78, 80

Lee, Steven P., 25n5
Lewis, David, viiin1
Liao, Matthew, 39n2

logic, laws of, 10, 22, 24, 28, 41, 42, 43, 44, 45, 49, 51, 52, 53, 54, 68, 70, 72, 82, 88, 92, 97, 100; logical duties, speaker's, 41, 42

Makkreel, Rudolf A., 95n1
Mann, Thomas, 17, 18
Mantzavinow C., 95n2
master speech acts, 2, 3, 4, 6, 7, 28, 35, 44, 57, 62, 65, 66, 68, 69, 72, 76, 101
memory bank, 10, 11, 12, 13, 15, 16, 17, 18, 21, 22, 23, 24, 27, 28, 31, 32, 35, 36, 37, 39, 46, 47, 50, 51, 52, 53, 54, 60, 61, 65, 66, 67, 68, 70, 73, 75, 76, 77, 78, 79, 80, 81, 82, 83, 84, 85, 87, 88, 89, 90, 92, 94, 99, 100, 101, 102, 103, 104; abstract range of, 11, 16, 23, 28; actual range of, 11, 16; common, 10, 11, 13, 15, 16, 24, 50, 80; general, 10, 11, 16; local, 9, 10, 11, 16, 18, 28, 30
memory bank annex, 88, 89, 90, 92, 94, 102, 103
memory bank dimensions: content, 12, 13, 15, 35; major speech act types, 12; range, 12, 16; source, 12; stability, 12, 16
memory bank helpers, 31, 32, 33, 67, 71, 100
Montague, Richard, viiin1
morality, common, 21, 22, 23, 27, 52, 84, 88, 99, 100, 102
moral thinking. *See* ethical thinking

Network, The, 28, 29, 30, 31, 78
Nocera, Joseph, 66, 67, 74n1
normativity, 17, 21, 24, 35, 37, 49, 52, 54, 59, 72, 82, 84, 101

Oulahan, Richard V., 63n2

Parfit, Derek, 25n3, 39, 39n3
perlocutionary acts, 94
presumptions (presuppositions). *See* context, linguistic

Quine, W. V., 97, 98, 99, 101, 104n1, 104n2

Scanlon, T. M., 25n2
Seanor, D., 8n6
Searle, John R., vii, viii, ixn4, 2, 4, 8n5, 8n6, 28, 29, 30, 31, 33n1, 33n2, 33n3, 33n4, 33n5, 59, 63n1
sincerity condition, 43, 66, 72
Shaw, William H., 25n8
speech act modifiers, 3
speech act stoppers, 3
speech act types: assertives, vii, viii, 3, 4, 15, 16, 17, 24, 27, 28, 29, 30, 35, 43, 49, 52, 53, 54, 59, 61, 62, 71, 72, 73, 77, 98, 99, 100, 101, 102; commissives, viii, 2, 3, 4, 28, 43, 59, 61, 73, 100, 101; declarations, viii, 28, 59, 60, 62, 100, 101; directives, viii, 2, 3, 4, 28, 43, 59, 62, 73, 75, 100, 101; evaluatives, viii, 2, 3, 8n6, 16, 28, 59, 62, 73, 100, 101; expressives, viii, 57, 58, 59, 61
speech activity, 2, 3, 4, 5, 7, 11, 14n1, 16, 23, 29, 35, 41, 43, 44, 51, 72, 78, 94, 97, 99
Stalnaker, Robert, viiin1, 10, 14n1
Strawson, P. F., 98, 104n3
Stakeholder Theory, 24

text, 1, 3, 5, 7, 9, 11, 12, 17, 18, 24, 28, 36, 41, 50, 51, 61, 71, 75, 76, 77, 78, 79, 80, 81, 87, 88, 89, 90, 92, 93, 94, 103, 104
theory construction, empirical, 36
theorizing, mid-level, 24
transcendental. *See* context, linguistic
Tuchman, Barbara, 17, 19n1, 19n2, 19n3

understanding, 91, 92, 94
utilitarian thinking, 24, 38, 82, 84

virtue theorizing, 38, 84

Walzer, Michael, 25n5

About the Author

Nicholas Fotion, emeritus professor of philosophy, Emory University, has written twelve books and more than one-hundred articles. *Theory vs. Anti-Theory in Ethics: A Misconceived Conflict* (2014) is his most recent book. His publications are primarily in the fields of ethics, ethical theory, medical ethics, military ethics, and the philosophy of language.

www.ingramcontent.com/pod-product-compliance
Lightning Source LLC
Chambersburg PA
CBHW030909040526
R18240000002B/R182400PG44116CBX00006B/5